Management for Professionals

More information about this series at http://www.springer.com/series/10101

Daniel Huber · Heiner Kaufmann
Martin Steinmann

Bridging the Innovation Gap

Blueprint for the Innovative Enterprise

 Springer

Daniel Huber
Mühlethurnen, Switzerland

Heiner Kaufmann
Münsingen, Switzerland

Martin Steinmann
Bern, Switzerland

Originally published in German with the title 'Bridging the Innovation Gap – Bauplan eines innovativen Unternehmens' by Springer Gabler in 2014.

ISSN 2192-8096   ISSN 2192-810X   (electronic)
Management for Professionals
ISBN 978-3-319-55497-6   ISBN 978-3-319-55498-3   (eBook)
DOI 10.1007/978-3-319-55498-3

Library of Congress Control Number: 2017939563

Cover illustration: eStudio Calamar, Berlin/Figueres

Printed on acid-free paper

This Springer imprint is published by Springer Nature
The registered company is Springer International Publishing AG
The registered company address is: Gewerbestrasse 11, 6330 Cham, Switzerland

Dedicated to our mentor, friend, and project benefactor, Christoph Gloor, who departed far too early from this world.

Foreword

Although many books have been written about innovation, there aren't many that offer any significant new insights into the subject. This book is one of the few that does. It is based on the many years the authors have spent examining the subject of innovation as it is approached in the literature and in real-world enterprise settings.

The "innovation gap" that the authors identify and very compellingly describe in the book is the reason why so many innovation projects are destined to fail. The book shows that the management models and approaches to innovation that are still applied in so many enterprises today are incomplete and to some extent misguided.

However, not content to merely point out the existing shortcomings of other approaches, the authors go on to present what, by my lights, is the first practicable, holistic model of innovation—a blueprint, as it were, that any enterprise can apply to develop a greater capacity for systematic innovation.

My own practical experience, gathered in the context of supporting hundreds of clients engaged in innovation projects on behalf of our innovation factory (Creaholic SA), confirms the relevance of the conclusions the authors reach in the book—particularly their suggestion that the laws of innovation are fundamentally different from those that are reflected in most conventional management approaches to innovation.

Creaholic SA   Elmar Mock
Biel, Switzerland

Acknowledgements

We would like to express our gratitude to the various organizations and individuals who made this book possible.

First and foremost, we would like to thank the Otto Wirz Foundation, the foundation's President, Götz Stein, and the foundation's Board of Directors. The Otto Wirz Foundation financed our book project and thereby established an important basis for the book. Special thanks go to two members of the Board of Directors, Heinrich Mühlemann and Cuno Wüthrich. They encouraged us to write the book and introduced the book project to the Board of Trustees.

A second organization that helped to make this book possible is the Management Center of the Department of Engineering and Information Technology at Bern University of Applied Sciences where the two coauthors Daniel Huber and Heiner Kaufmann are employed. Much of what is in this book is based on results that were worked out at the Management Center. Our special thanks in this context go to Christoph Gloor and Arno Schmidhauser.

Christoph Gloor, the deceased former Director of the Management Center, was a major supporter of our project and, without a doubt, our most important mentor when it came to the subject of innovation. He was the one who encouraged us at a very early stage to undertake the project. He personally convinced many of those who were involved of the project's importance. He was also especially generous with his time and energy, taking on a lot of additional work to give us an opportunity to concentrate on the project.

We also owe a debt of gratitude to Arno Schmidhauser. As Christoph Gloor's supervisor, he gave his support to the project and did his part to make sure we had the time we would need to make progress.

Thanks are also due to the employees of the administrative team at the Management Center who were often forced to do without their boss and who nonetheless did an excellent job keeping the university department up and running.

We also wish to thank Firmament AG, Martin Steinmann's company,[1] for its valuable contribution to the quality of the book. It is indeed a great sacrifice for a

---

[1]Firmament AG, Monbijoustrasse 27, CH-3011 Berne, Switzerland.

small enterprise to do without such a crucial employee to the extent that it did. We are extraordinarily grateful for this setting of priorities.

Our thanks also go to the employees of InoBooster,[2] an enterprise operated jointly by Martin Steinmann and Daniel Huber. The extensive discussions on innovation that we carried out in this context helped us to focus our inquiry and provided a means of continuously verifying our results with reference to real-world cases.

We also wish to thank Giuliano Guscelli at innoBE AG[3] for his valuable contribution to the development of the "Innovative Enterprise" training program for enterprises.[4] Many of the ideas we discussed in this connection appear in the third part of the book.

The Springer International Publishing AG naturally also played a crucial role in the English edition of this book. Our special thanks in this regard go out to our Editor Dr. Prashanth Mahagaonkar and his Senior Editorial Assistant Ruth Milewski who accompanied our project in an excellent and highly professional manner. Working together with them was an extraordinarily constructive and pleasant experience.

We are indebted to Elmar Mock, the designer of the Swatch watch and founder of the company known as Creaholic SA,[5] for many fruitful discussions on innovation in recent years and particularly for agreeing to write the foreword to the book.

Last but not least, our thanks go to our wives, companions, and families. They took our decision to devote so many hours of our free time to writing the book with composure, kindness, and support.

---

[2]http://www.inobooster.com

[3]https://be-advanced.ch

[4]https://www.ti.bfh.ch/de/weiterbildung/dienstleistungen/angebote_fuer_unternehmen/innovative_enterprise.html

[5]http://www.creaholic.com

Contents

About the Authors

Daniel Huber is the Director of the Management Center at Bern University of Applied Sciences, Professor of Innovation Management, and Director of the University's Executive MBA program in Innovation Management. Daniel Huber has directed the development of the program in innovation since 2008. He graduated from the Swiss Federal Institute of Technology (ETH) in Zurich with an advanced degree in Engineering and completed an advanced degree in Management at the International Institute for Management Development (IMD) in Lausanne. Daniel Huber worked for more than 20 years at Swisscom's innovation division. During this time, he gained an in-depth understanding of innovation as it is practiced in large enterprises. He has many years of experience in various areas of innovation management. Working together with two other managers, he was responsible for setting up Swisscom Innovations, an innovation unit with around 180 employees. This gave Daniel Huber an opportunity to acquire an extensive knowledge of innovation management under optimal conditions. Daniel Huber is a member of the Board of Directors of swissfuture, the Swiss association of futurologists.

Heiner Kaufmann is a Professor of Business Creation at the Management Center of Bern University of Applied Sciences and Director of the University's Executive MBA program in Innovative Business Creation. He completed academic programs at the Swiss Federal Institute of Technology (ETH) in Lausanne and at the Massachusetts Institute of Technology (MIT) in Boston, USA. Heiner Kaufmann worked for almost 20 years as a manager and project manager in a number of high-tech companies, particularly in the areas of innovation and research and development. He is the founder of a start-up enterprise and an innovation consulting company that was acquired by the innovation factory Creaholic SA in 2012. In recent years, Heiner Kaufmann succeeded in developing a new learning and teaching concept for business conception.

 **Martin Steinmann** has been a successful Business Enabler in the area of information and communications technology (ICT) for more than 25 years. He has positioned himself as a skilled interpreter at the interface between technology and business. Steinmann has also been an independent enterprise consultant in the area of strategy and innovation management since 2002. His clients include prominent major corporations and numerous ambitious medium-sized enterprises. Prior to his current work, he was employed in various management-level positions at a number of international companies. Martin Steinmann has a degree in Economics and Business Administration from the University of Fribourg (International Institute of Management in Telecommunication). With his extensive experience, he has gained an excellent reputation as a solution designer and coach, especially in start-up situations. In addition to this, he works as an expert and thesis adviser for Master's theses in EMBA in Innovation Management at the Management Center of Bern University of Applied Sciences.

List of Abbreviations

BFH	Bern University of Applied Sciences
CEO	Chief Executive Officer
EMBA	Executive Master of Business Administration
ETH	Swiss Federal Institute of Technology
IPR	Intellectual property rights
MBTI	Myers–Briggs type indicator
MVP	Minimal viable product
MZBE	Management Center at Bern University of Applied Sciences
NT	Intuition, thinking (personality type as per MBTI)
OWS	Otto Wirz Foundation
SJ	Sensor, judging (personality type as per MBTI)
TIME	Telecommunications, information, media, and entertainment

List of Figures

List of Tables

The Origins

1

The present book is the product of a personal examination of innovation that began more than 30 years ago. When I look back at my professional career, it occurs to me that I have concerned myself almost exclusively with the subject of innovation ever since I enrolled as a student at the Swiss Federal Institute of Technology in Zurich. While this was not what I expressly set out to do—by design, as it were—it was also not simply a matter of chance. After all, I am affected myself to some extent by the subject described in the book too and by the findings that have resulted from my examination of it.

I spent the first 20 years of my professional career in a corporate setting. As I've already mentioned, I concerned myself during this long period nearly exclusively with the subject of innovation—in all of its various forms and at all levels of corporate management. As described in greater detail in what follows, I had the good fortune of experiencing a period of well-functioning and systematic innovation. It wasn't until much later that it dawned on me how extraordinarily rare such developments are. While pursuing my work, I had merely often wondered why a few projects seemed to develop splendidly while so many others failed. At the time, however, I would not have been able to explain what I had experienced.

Years later after accepting a position at the university, I was assigned to teach a course in innovation management. To my surprise, I soon discovered that there were no suitable textbooks that addressed all of the topics I regarded as essential for a course on innovation. Moreover, most of the explanatory material for the processes and events I had so often experienced in the corporate world seemed woefully inadequate. I was therefore forced to compile my own material so as to account for all of the elements that seemed important to me and that I knew from the literature and my experience in enterprise settings.

With such a comprehensive and integrated approach I thought that, ideally, I would also essentially be able to explain my own experience. This aim served then

---

All illustrations are published with the kind permission of © Heiner Kaufmann, Daniel Huber, and Martin Steinmann. All Rights Reserved.

© Springer International Publishing AG 2017   1
D. Huber et al., *Bridging the Innovation Gap*, Management for Professionals,
DOI 10.1007/978-3-319-55498-3_1

as test in the context of preparing for the course, by which to assess the completeness and quality of the course design and content. At the same time, I knew that innovation theory was still a very young and thus incomplete science and it would not be very probable that I would reach my goal.[1]

This intensive preparatory phase at the university turned out to be a phase of intensive integration work. In order to arrive at the big picture of the subject, the many individual, and to some extent self-contained, findings described in the literature needed to be integrated for the first time. To my surprise, many of the rules of thumb that had proven their mettle in corporate settings received no mention whatsoever in the literature. Moreover, the process of integrating the many different sources led to further important insights. Altogether, it is fair to say that I was now evaluating many years "corporate-level experimentation" in the area of innovation management (my many years of hands-on, corporate-level experience) drawing upon it as a source for establishing an adequate theoretical perspective. And of course my efforts were naturally accompanied by a vigorous exchange of ideas with the many innovation specialists and managers in my professional network.

The result was a series of lectures in innovation management that is still held in the framework of an Executive MBA (EMBA) program offered at the Management Center of Bern University of Applied Sciences for its approximately 300 enrolled students. The course has now been offered every semester for more than 8 years.

EMBA students typically hold key positions of responsibility at various enterprises and complete their studies on a work-study basis. They are typically between 30 and 50 years old and are capable sharing their extensive professional experience in the context of completing their course work. It was therefore not only possible to reconcile and refine my newly won insights with my own professional experience, but also with the professional experience of a large number of EMBA students. To my own surprise, I was ultimately able to largely explain my own experiences in terms of a well-developed, comprehensive system. The new insights therefore had passed the test.

The continuing success of the innovation management course then provided a rationale for documenting and publishing the results. With the generous support of the Otto Wirz Foundation and my employer, Bern University of Applied Sciences, precisely this has been accomplished in the form of the present book.

It would not have been possible to realize the book in its present form without the valuable assistance of my two co-authors, Heiner Kaufmann from Bern University of Applied Sciences[2] and Martin Steinmann from InoBooster[3] and Firmament AG.[4]

---

[1]Innovation theory remains a young and incomplete discipline.

[2]www.mzbe.ch,   https://www.ti.bfh.ch/de/weiterbildung.html   or   https://www.ti.bfh.ch/de/weiterbildung/weiterbildungsangebote/emba/innovative_business_creation.html

[3]www.inobooster.com

[4]Firmament AG, Monbijoustrasse 27, CH-3011 Berne, Switzerland.

Heiner Kaufmann and I have collaborated on various projects throughout the years. Much of this work is described in detail in another book entitled "Innovation Factory" (Mock et al. 2013), which offers the reader a detailed account of Heiner Kaufmann's second employer, Creaholic AG,[5] a genuinely unique innovation factory.

The present book is indebted to Heiner Kaufmann for numerous diagrams and illustrations, as well as a valuable description of the exploration process. Indeed, the graphic illustrations of various key concepts give the book greater clarity and accessibility.

Martin Steinmann has also accompanied me for many years in various academic and professional capacities. In addition to running his own consulting firm, Firmament AG,[6] Martin Steinmann is a partner and co-founder of our jointly owned innovation company known as InoBooster.[7] The aim of InoBooster is to make the insights described in this book available for real-world application in public and private enterprises. The present book would scarcely have reached its current degree of maturity without the numerous and extensive discussions with Martin Steinmann. Steinmann made major contributions to the precise formulation and use of terms and definitions, as well as to their grounding in the practical world of innovation management. Beyond this, Steinmann contributed the lion's share of the examples that appear in the book.

With its unconventional approach to the subject of innovation, namely, the deriving of a coherent, holistic theory based on extensive practical experience, the present book succeeds to introduce for the first time a complete and integrated model of innovation. The suitability of this new model for practical application has therefore already been extensively confirmed in practice and does not, as is normally the case, need still to be demonstrated. Indeed, this new innovation model has enabled me to explain my own practical experiences.

1.1 Keys to Successful Innovation: A Summary

In recent years, more and more market observers have come to emphasize the role of innovation in the success of enterprises. This is also reflected in the increased attention the subject has received in the literature. However, despite the widespread agreement on the centrality of innovation, most enterprises still are struggling with the topic of innovation and regularly experience great difficulties in their innovation efforts. On the contrary, facing the issue at all can feel a lot like facing a dilemma. Consider the words of Henry Chesbrough, *"Most innovations fail. And companies that don't innovate die"* (Chesbrough 2003, p. xvii). Drawing on our extensive experience in enterprise settings, it is precisely this dilemma that we seek to solve in the present book. In approaching this challenge, we base our analysis on a period of practical innovation during which systematic efforts to achieve innovation over the course of many years were ultimately crowned with success. Given that successful

---

[5] www.creaholic.com

[6] Firmament AG, Monbijoustrasse 27, CH-3011 Berne, Switzerland.

[7] www.inobooster.com

attempts to innovate are far more the exception than the rule, we decided to conduct an in-depth examination of the case at hand in the hopes of distilling out the essential components of a successful approach to innovation. Our findings are presented throughout the present book. The purpose of the present summary is to provide a condensed account of the most important findings.

It turns out that the innovation process, as it has been described up until now, is incomplete. It is not primarily the ideation phase and the idea selection phase that precede the development phase, but the new and more complex **early warning** and **exploration** phases. In particular, the tendency to conceive of the goal as the presentation of a "selection of best ideas" for subsequent implementation is a misleading oversimplification of the task at hand.

The early-warning-system phase encompasses the idea generation function as already depicted in the innovation process up to now. But this function is now newly complemented by a new function of systematic gathering of information.

In contrast, the exploration phase is entirely new. The exploration phase is essentially a matter of a complex procedure in which the identified new ideas are joined as conceptual building blocks to other—largely preexisting—building blocks to form an entirely new composite system. In doing so, explorers take account not only of the technological building blocks, but also the commercial building blocks, including the various elements of the business model and the relevant business context. This allows one to arrive at a reasonably reliable estimate of the commercial value of a potential innovation before moving forward to the costly development phase. As other authors have pointed out, it is the business context that determines the value of an innovation. In contrast, it is not possible to assign a commercial value to ideas alone.

Figure 1.1 offers a graphic illustration of the exploration phase.

The **exploration arena** (shown on the left in the graphic) represents the introduction of the conceptual building blocks identified in the early warning system, building blocks that are then refined in the four processing steps: Characterize, Combine, Optimize and Complete (zones 1–4) to form substantial business opportunities. These business opportunities are then prepared for purposes of

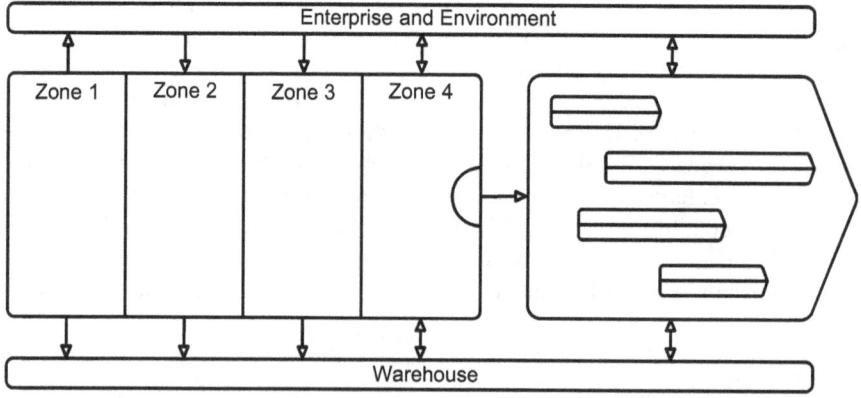

Fig. 1.1 Exploration

decision making in the framework of **exploration projects** (right). It is not until this phase that the innovation project can be assigned a commercial value.

One further result of our examination is the following realization: To handle the two early phases of the innovation process that we refer to as the early-warning-system and exploration phase, it is vitally important to assign individuals of a specific personality type, i.e. a type that differs from that of those individuals typically responsible in the enterprise for standard operational activities. This finding has as a consequence that enterprises that are interested in enhancing their capacity to innovate will have to accommodate two different enterprise cultures within the same organization. This requirement however is diametrically opposed to a number of broadly accepted organizational principles. This has a number of far-reaching consequences, and is also the reason why innovation presents enterprises with such a stark dilemma. Indeed, the necessity of accommodating two different enterprise cultures is probably the most important of the conclusions we arrive at in the present book.

A degree of misunderstanding and conflict is unavoidable at the points of contact between the two enterprise cultures. Moreover, such conflicts can lead to the systematic failure of innovation projects. We refer to these points of conflict collectively as the **innovation gap**. In order to bridge this innovation gap, we introduce an additional phase to the innovation process that we refer to as the transfer phase.

The complete innovation process can be represented as shown in Fig. 1.2.

Various organizational consequences arise from the necessity of accommodating two different enterprise cultures within an innovative enterprise. Those responsible for the two early innovation phases "early warning system" and "exploration" will need to occupy a central position in the enterprise and should report directly to the CEO. They are also to remain in close contact with those responsible for strategy

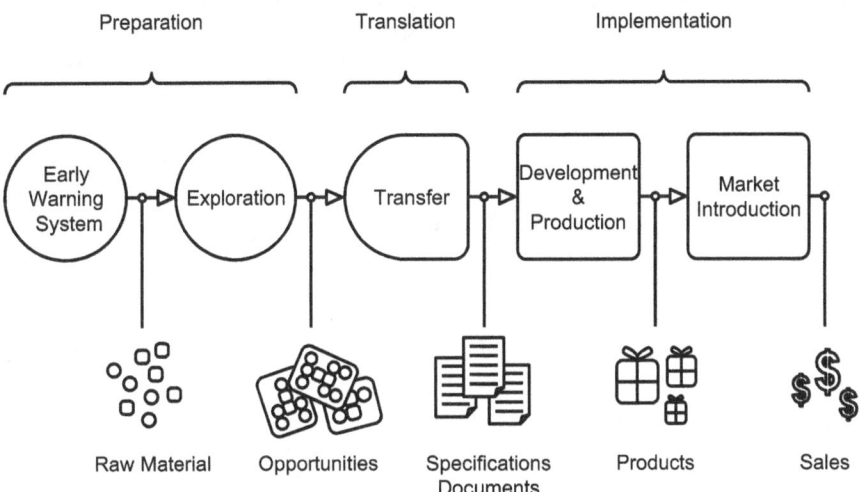

Fig. 1.2 The innovation process

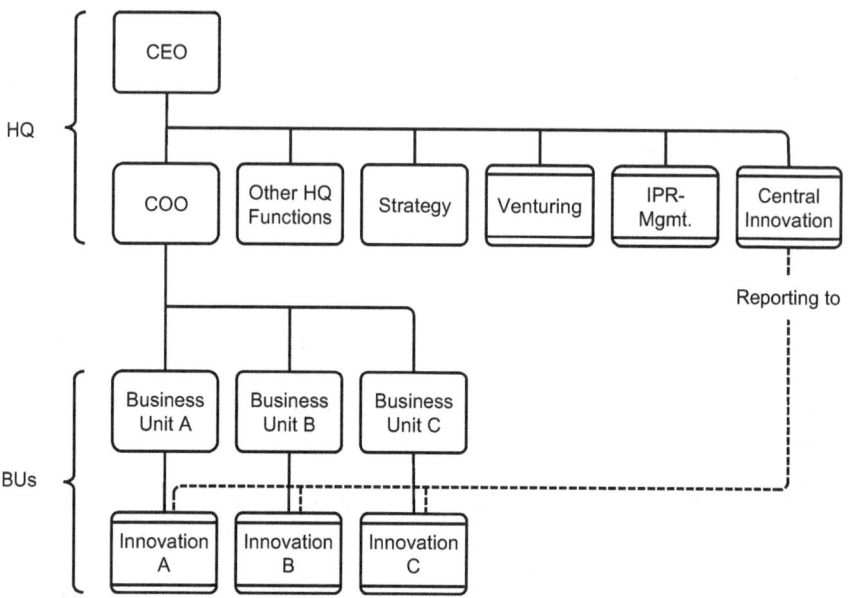

Fig. 1.3 Model organizational structure

development. Moreover, it will also be necessary to introduce the two new roles which we refer to as "venturing" and "IPR management."

The enterprise's new organizational structure can be represented as shown in Fig. 1.3.

Another consequence of the two different enterprise cultures arises in connection with the management and the system of incentives used for the two different employee types. Here, it is important to bear in mind that the employees responsible for the early-warning-system and exploration phases will have to be managed according to completely different principles than those used in the case of other employees in the enterprise. It is especially important in this context to consider the **leadership triangle** shown in Fig. 1.4.

Finally, successful innovation will also depend on well-calibrated strategies. Our experience in this connection shows that the **dual strategy approach** espoused by Derek Abell is especially suitable. As a supplement to the classic enterprise strategy that regulates what needs to be done today to secure today's business, Abell's approach takes account of a second strategy to regulate what needs to be done today to secure tomorrow's business. Abell refers to this second strategy as the "today-for-tomorrow strategy." This second strategy essentially encompasses the contents of the innovation strategy, which is thereby transformed from a functional, supplementary strategy to a fully equal part of the new dual strategy.

At the end of the book, we combine all of these findings and recommendations into the **Bern Innovation Model** (see Fig. 1.5) and demonstrate how and in what order this model can be implemented in the real world.

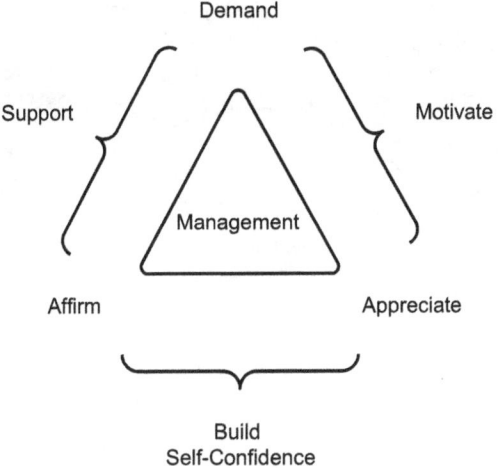

Fig. 1.4 The leadership triangle

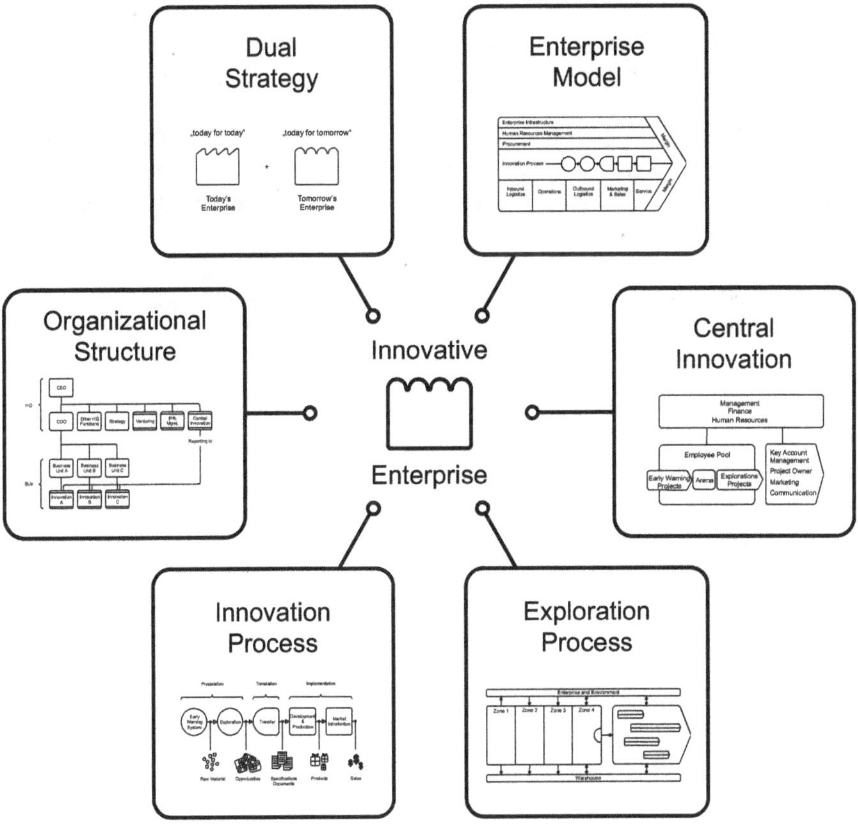

Fig. 1.5 The Bern innovation model

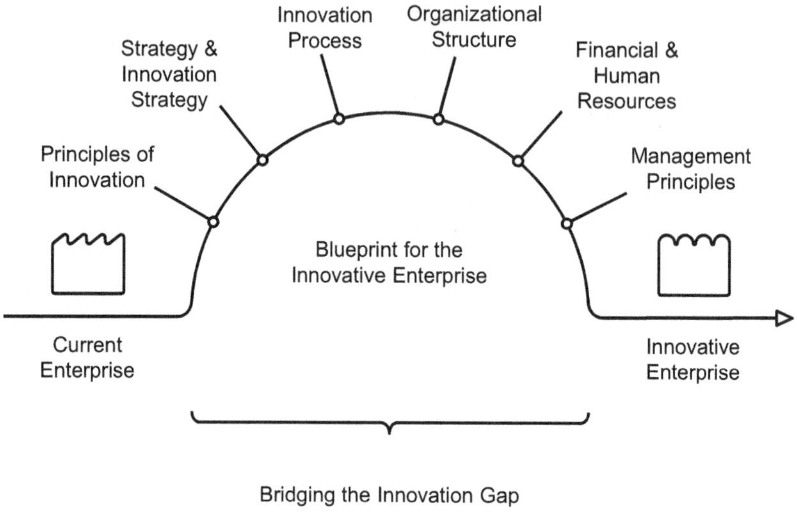

Bridging the Innovation Gap

Fig. 1.6 Blueprint of the innovative enterprise

The existing enterprise is thereby transformed into an enterprise with an enhanced capacity for innovation. This transformation takes place in six steps, as shown in Fig. 1.6.

While systematic approaches to innovation can be successful, they require the application of a new set of enterprise rules. This does not mean, however, that it is possible to eliminate the risk associated with innovation projects. A degree of risk will always remain.

References

Chesbrough, H. (2003). *Open innovation*. Boston: Harvard Business School Press.

Mock, E., Garel, G., Huber, D., & Kaufmann, H. (2013). *Innovation factory*. Fribourg: Growth Publisher.

Part I

The Problem

Innovation: An Abiding Enigma

2

Innovation is a dilemma.

> *Most innovations fail. And companies that don't innovate die.*
> (Henry Chesbrough)

Innovation is the talk of the town. It is perceived by many in the corporate world as a silver bullet that will secure sustainable growth and long-term prosperity. And in many cases, it is exactly that. Numerous studies[1] and prominent success stories give testimony to the positive effects of successful innovation. Today, innovation is thought of as the most important generator of solid economic development and wealth creation. The fact that innovations can have a sustainable impact on enterprise development is hardly surprising given that it is all about shaping the future of the enterprise.

▶ Innovation is all about shaping the future of the enterprise.

2.1 Innovation: A Dilemma

Despite the success stories, it is obvious that many attempts to innovate end in failure. Even specialists in the area of innovation are dissatisfied when it turns out that not even one of ten attempts to innovate proves successful. As shown in Fig. 2.1, successful innovation projects represent the small tip of an iceberg predominately comprised of failed innovation projects.

All illustrations are published with the kind permission of © Heiner Kaufmann, Daniel Huber, and Martin Steinmann. All Rights Reserved.

[1]For instance: McKinsey study "Entrepreneurship in Germany," published in Meffert and Klein (2007).

© Springer International Publishing AG 2017
D. Huber et al., *Bridging the Innovation Gap*, Management for Professionals,
DOI 10.1007/978-3-319-55498-3_2

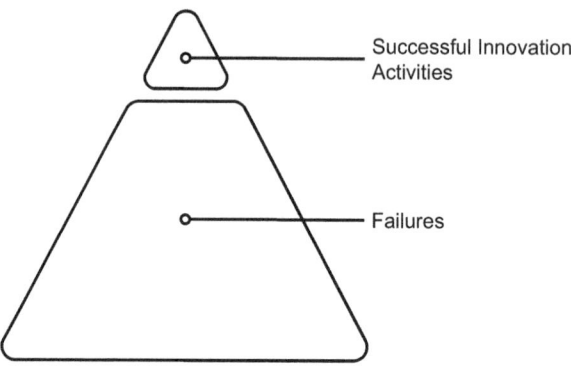

Fig. 2.1 The innovation iceberg

Enterprise managers are often well-acquainted with innovation gridlock or out-right failure. *"Most innovations fail. And companies that don't innovate die,"* says Henry Chesbrough (2003, p. xvii). This quote gives expression to the conflicts faced by enterprises eager to secure the benefits of innovation. On the one hand, innovation appears to be essential for their well-being and survival. On the other hand, attempts to innovate are fraught with risk and can therefore threaten their very existence. Failure is indeed a typical outcome of attempts to innovate. The question therefore arises as to the extent to which this quandary is simply a part of the game or whether the usual attempts are marked by some, as yet poorly defined, albeit fundamental flaws.

2.2   The Crux of the Problem

The difficulties associated with attempts to innovate, as well as the growing consensus on the necessity of innovation, have inspired extensive discussions on the subject of innovation in professional journals. These contributions to the literature include definitions of various types of innovation, extensive study results and various rules and recommendations to be mindful of when developing innovation projects. While all of these contributions are valuable, they tend to shed light only on specific aspects of innovation and have not yet been integrated into a generally accepted or standard model. In the area of innovation management, there is still no satisfactory systematic approach to innovation, let alone a specific and comprehensive methodology for significantly increasing the likelihood of successful innovation.

The problem therefore is that a very high percentage of innovation projects fail that we still do not know what to do to improve the prospects for success.

2.3   Innovation: A Difficult Definition

The difficulties seem to crop up as soon as we attempt to define the term *innovation*. Authors have so far tended to formulate their own definitions of innovation. Fortunately, a more uniform interpretation of the term has gained acceptance in recent years. In addition to the properties of novelty and relevance, this new

Fig. 2.2 The essential
elements of innovation

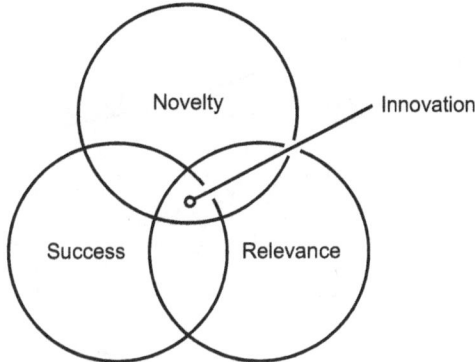

interpretation takes account of the element of success (e.g. market success). In Fig. 2.2, we use a Venn diagram to clarify this interpretation of the term innovation, with innovation constituting the area of convergence of the elements novelty, success and relevance.

Consider, for instance, the Wikipedia definition:

> Innovation is the creation of better or more effective products, processes, services, technologies, or ideas that are accepted by markets, governments, and society.[2]

This definition of the term innovation includes success. This proves to be very useful because it clarifies that innovation is not something that happens in enterprises, but something that takes place in the area of application, often on the market itself. Unsuccessful attempts to create innovations therefore do not qualify as innovations. This helps to clarify the term for purposes discussion.

2.4 Consulting the Literature for Clues to Innovation

What do the experts have to say about achieving innovation? While the advice one finds in the literature and in textbooks varies considerably with respect to many details, there do seem to be a number of common assumptions:

- Innovation is necessary for the long-term prosperity of enterprises.
- Innovation projects should take place outside the framework of daily business.
- It's always an organization, usually an enterprise, that is responsible for driving and realizing innovation.
- There is always an innovation process that takes place within this organization.
- Various specific organizational forms are proposed as being more suitable for generating successful innovation than others.
- A favorable enterprise culture is important.

---

[2]http://en.wikipedia.org/wiki/Innovation, reference date: December 28, 2011.

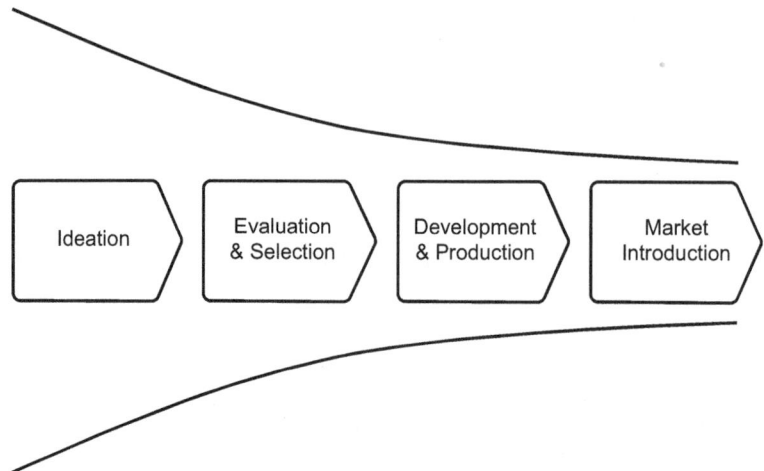

Fig. 2.3 The funnel model of the innovation process

While many different proposals have been made about the most appropriate organizational structures and enterprise cultures for achieving innovation, there is widespread agreement in the literature about the structure of the innovation process. As shown in Fig. 2.3, the innovation process is typically represented using a funnel model (e.g. Müller and Dörr 2011, p. 31).

According to textbook accounts, the innovation process begins with an ideation phase in which a large number of ideas are generated.[3] The literature offers a variety of methods for approaching the actual generation of ideas during this initial phase.[4] The generated ideas are then evaluated (cf. Eversheim 2003, p. 93)[5] and the very best are selected for purposes of development (e.g. a new product). After successful development, the novel products are then produced and subsequently introduced to the market. If, as indicated above, their market introduction is met with success, then the products in question qualify as genuine innovations.

In general, one can assume that the number of innovation projects that are pursued simultaneously will tend to continuously decrease from process phase to process phase. This is illustrated in the image of the funnel. If this decrease is explicit in the evaluation-and-selection phase, it will tend to be implicit in all subsequent phases. There is, on the one hand, the awareness that the likelihood of project failure is relatively high. Added to this is the fact that the subsequent project phases become more complex and costly. And most enterprises do not have the necessary resources to keep all that many late-stage innovation projects on life support.

---

[3]"To have a great idea, have a lot of them" (Thomas Alva Edison).

[4]E.g. via brainstorming. A collection of different methods of generating ideas is available, for instance, in Vahs and Brem (2013), p. 280 ff.

[5]E.g. via cost-benefit analysis.

When comparing the contributions made in the literature, we find that while all of the authors in question essentially describe the same funnel model, they differ in terms of their focus and the degree to which they provide detailed accounts of specific phases depending on the orientation of their publication and the industry whose perspective they are depicting (Vahs and Brem 2013, p. 231 ff.). For instance, the development phase is often variously subdivided into a specifications-document phase and a pure development phase. The same can be seen when it comes to the production and market-introduction phases (see Vahs and Brem 2013, p. 232 ff.).

This does not apply, however, to the first two phases, the ideation and evaluation-and-selection phases. It is evident that these two phases are currently less amenable to proper structuring. While many different methods are available for the generation of ideas (e.g. Vahs and Brem 2013, p. 278 ff.), published accounts of the evaluation phase do not yet go much beyond cost-benefit analyses (Eversheim 2003, p. 93). Owing to the existing lack of clarity, these first two phases are often referred to as fuzzy front-end phases (see Boeddrich 2004 or Kahn 2012, p. 450).

Since the 1980s, however, considerable progress has been made in the area of innovation management. Many innovation management mysteries, for instance, have been explained. This progress includes the following contributions:

- Giovanni Dosi's work on technological paradigms and technology trajectories (Dosi 1982).
- William Abernathy and Kim Clark's distinction between architectural, market-niche, regular and revolutionary innovations (Abernathy and Clark 1985).
- Rebecca Henderson and Kim Clark draw a distinction between incremental, modular, architectural and radical innovations (Henderson and Clark 1990).
- In his book *Open Innovation*, Henry Chesbrough presents fundamentally new concepts, particularly with regard to the exploration part of the innovation process (Chesbrough 2003).
- W. Chan Kim and Renée Mauborgne place strategic aspects of innovation activities in the foreground and deliver the strategic contour as a suitable method (Chan and Mauborgne 2005).
- In a number of books published in the 1990s, Geoffrey Moore explained how to introduce innovations to the market and to avoid common pitfalls while doing so (e.g. Moore 1995).
- In his book *The Innovator's Dilemma*, Clayton Christensen presents explanations for a number of surprising innovation successes (and a number of no less surprising failures) and introduces a new class of innovations referred to as *disruptive innovations* (Christensen 1997). Disruptive innovations are capable of forcing entire markets to undergo processes of restructuring and sometimes lead to the dethroning of the originally dominant market players.
- Steve Blank examines the development process of startups and introduces the customer-development process (Blank and Dorf 2012, p. 19). Steve Blank also points out the fundamental difference between a startup and a normal enterprise,

with startups being organizations that are on the lookout for a business model while enterprises are organizations that have already found a business model.

- Building upon the work of Steve Blank, Eric Ries introduces the concept of the lean startup and the minimum viable product (Ries 2011, p. 77).

Despite these contributions, however, innovation management remains an extraordinarily difficult discipline.

2.5   Assessing the Status of Innovation in Practice

"Most innovations fail. But companies that don't innovate die" (Chesbrough 2003, p. xvii). Are we to take this statement at face value, or was Henry Chesbrough just making a point? After all, there seems to be no end to (successful) innovations.[6] Perhaps the management of innovations isn't as bad as it's made out to be. Let's consider the situation from the perspective of the results.

A look at the innovation results reveals the following facts: Yes, there is no end to (successful) innovations. And when we analyze the individual success stories, we find the following recurring pattern:

- The corresponding innovation projects were abandoned (often on multiple occasions) before they became successful.
- After their official abandonment, however, they were often continued, against the rules, in an unofficial capacity and then later resumed as official projects . . . before, oftentimes, being scrapped yet again. Depending on the case, this pattern may include multiple repetitions.
- The innovations were nearly always successful because one or only a very few specific persons. If the involvement of these key individuals had been removed from the innovation projects, there would very likely have been no success (see Mock et al. 2013, Chap. 2, p. 19 ff.).

It turns out that most (successful) innovations depend on particular individuals and that their development often proceeds despite official enterprise decisions to cut off their support! This offers a strong indication that existing systematic innovation management programs in enterprises are not really working the way they should be.

▶   Successful innovations usually depend on particular individuals and often
      arise despite official enterprise decisions to cut off their support.

In light of these findings, it is instructive to take a closer look at the current innovation practices of enterprises. Most enterprises maintain officially approved innovation programs. Larger enterprises will often even have taken the further step

---

[6]There are also very successful copy cats. While copy cats represent a very specific strategy of exploiting the market potential associated with the innovations of others, they can be highly successful. All that is required is an established brand and a critical mass.

of establishing independent organizational units to cover innovation. In many cases, the enterprises are proud of their innovation efforts and are accordingly willing to invest considerable amounts of money in them. Our investigations, however, indicate that this typical approach to innovation seldom leads to commercial success. Many innovation projects fail at the point of market introduction, and far more fail before any sort of new product is ever readied for a market launch (Vahs and Brem 2013, p. 54). This means that most development projects are abandoned or are simply never brought to a successful conclusion. And very many more good ideas never even make it to the development phase.

To find out more, we can ask the enterprise executives themselves whether they are satisfied with the results of the innovation programs that have been initiated in their enterprises. When we do this, we are usually told that the program results have fallen way short of expectations and that the initiated innovation projects have tended to lack efficiency. There is a prevailing atmosphere of impatience and a pronounced desire for the enterprise to at last take a fast track to innovation (see Dueck 2013, p. 280; BCG 2010; McKinsey 2008). However, when asked how this is to be managed, the responses reveal a sense of helplessness. This strongly suggests that enterprise executives are also of the opinion that their internal innovation management programs are not really doing what they are supposed to do.

All things considered, we wind up with grounds for concern: Only a small percentage of the innovation ideas that are assessed as promising actually turn out to be successful. And the lion's share of the few ideas that make it owe their success to the determination of individuals who are forced to proceed without the support of their own enterprise management. The success of enterprise-approved and systematically managed innovation projects ranges from disappointing to non-existent!

2.6   Conclusion and Proposition 1

The success of most systematic innovation efforts is apparently marginal while cases of greater success seem to depend on the efforts of particular individuals and often come despite explicit decisions on the part of enterprise management to cut off their support. We can conclude from this that there remains a broad lack of understanding as to how to improve an enterprise's capacity for (successful) innovation.

▶   Proposition 1: Innovation has remained an enigma.

The Research Question: What Is the Best Approach to Systematic Innovation?
Based on our Proposition 1, that innovation has largely remained an enigma, and Henry Chesbrough's suggestion that most innovation projects fail, we will now attempt to identify the reasons for the failure. After all, if you know why systematic approaches to innovation fail, you should be able to show the prerequisites that need to be fulfilled for the ability to carry out successful systematic innovation. This second point is naturally of crucial significance for economic development. Our first question as to why most innovation projects fail is therefore essentially subordinate to our second question as to the best approach to systematic innovation. Our research question can therefore be formulated as follows:

What is the best approach to systematic innovation? Questions subordinate to this one include:

- Why do most innovation projects fail?
- What model do enterprises need to follow to achieve successful systematic innovation?
- How can this be achieved?

References

Abernathy, W., & Clark, K. (1985). Innovation: Mapping the winds of creative destruction. *Research Policy, 14*, 3–22.

Blank, S., & Dorf, B. (2012). *The startup owner's manual*. Pescadero, CA: K&S Ranch.

Boeddrich, H.-J. (2004). Ideas in the workplace: A new approach towards organizing the fuzzy front end of the innovation process. *Creativity and Innovation Management, 13*(4), 274–285.

Boston Consulting Group (BCG). (2010). *Innovation 2010*. Boston: Consulting Group Report.

Chan, K. W., & Mauborgne, R. (2005). *Blue Ocean strategy*. Boston: Harvard Business School Press.

Chesbrough, H. (2003). *Open innovation*. Boston: Harvard Business School Press.

Christensen, C. M. (1997). *The innovator's Dilemma*. New York: Harper Business.

Dosi, G. (1982). Technological paradigms and technological trajectories: A suggested interpretation of the determinants and directions of technical change. *Research Policy, 11*(3), 147–162.

Dueck, G. (2013). *Das neue und seine Feinde*. Frankfurt: Campus.

Eversheim W (Hrsg) (2003) Innovationsmanagement für technische Produkte. Springer, Berlin

Henderson, R., & Clark, K. (1990). Architectural innovation: The reconfiguration of existing product technologies and the failure of established firms. *Administrative Science Quarterly, 15*(1), 9–30.

Kahn, K. B. (2012). *The PDMA handbook of new product development*. Hoboken: Wiley.

McKinsey. (2008). Leadership and innovation. *The McKinsey Quarterly*, 1. http://www.mckinsey.com/insights/innovation/leadership_and_innovation. Reference Date: September 17, 2014.

Meffert, J., & Klein, H. (2007). *DNS der Weltmarktführer*. Heidelberg: Redline Wirtschaft.

Mock, E., Garel, G., Huber, D., & Kaufmann, H. (2013). *Innovation factory*. Fribourg: Growth Publisher.

Moore, G. A. (1995). *Inside the Tornado*. New York: Harper Business.
Müller, T., & Dörr, N. (2011). *Innovationsmanagement*. Munich: Hanser.
Ries, E. (2011). *The Lean startup*. New York: Crown Business.
Vahs, D., & Brem, A. (2013). *Innovationsmanagement*. Stuttgart: Schäffer-Poeschel Verlag.

The Missing Link: The Innovation Gap

3

The aim of the present chapter is to show what is lacking in the current state of the innovation theory, what the missing link in the chain is. If our hypothesis is right, i.e. that essential parts of the innovation process are indeed missing and that this is the reason why innovation does not work satisfactorily in practice, then we can expect that most innovation projects will terminate at some point in the process. It would be convenient in the context of diagnosing the problem if such points of termination were to turn up with greater frequency at specific locations in the process. This could give us an important clue as to the location in the innovation process that has not yet been sufficiently described.

3.1 What Is Missing for Successful Innovation?

Let us first have a look at how the textbook case of innovation is supposed to arise. As described in Chap. 2, the prevailing view in the literature is that the innovation procedure can be best represented by the funnel model, i.e. the innovation process consists of the following four phases: ideation, evaluation and selection, development and production and market introduction.

Let us therefore look for locations at which the innovation process typically terminates in practice. While doing so however it is important to bear in mind that innovation projects may fail during any of the phases of the process. It is even probable that many such projects will fail at some point because we cannot expect that all of the ideas submitted for consideration will at some point lead to success. Indeed, limited resources alone will force one to whittle down the prospective innovations. On the other hand, an ideal innovation process would have to be

---

All illustrations are published with the kind permission of © Heiner Kaufmann, Daniel Huber, and Martin Steinmann. All Rights Reserved.

configured so as to ensure that the ultimate selection can be made explicitly in keeping with certain milestones and on the basis of a clear set of criteria. It follows that if we want to examine the innovation process according to weaknesses and missing elements, then it will be primarily important for starters to identify implicit and unwanted process terminations.

At What Locations Do Innovations Typically Fail? What Do the Innovation Practitioners Say?

Let's review the four process phases of the classic funnel model of innovation and have a look at how successful the respective phases are in practice.

- Ideation phase: Various approaches to ideation are in use today. Depending on the particular enterprise, these approaches may be more or less haphazard, opportunistic or systematic, and may involve the use of a wide array of creativity techniques. It is very seldom the case that the innovation process is grounded during this initial phase. More often, we see hasty decisions to embrace a particular innovation idea, i.e. without first having considered the viable alternatives. This can lead to a situation in which opportunities for innovation are missed or in which innovation projects are not properly aligned to market developments. Despite possible conceptual problems, project termination is unlikely during this first phase.
- Evaluation and selection phase: This is the phase in which the ideas that are generated in the ideation phase are to be evaluated and whittled down to a manageable number for consideration and further processing in subsequent phases. This is usually done on the basis of a cost-benefit analysis. While this approach is generally to be called into question, as we will see below, any decisions to terminate a project on the basis of a cost-benefit analysis will at least tend to be explicit and transparent. In other words, such project terminations are not a matter of what we have referred to as unwanted, implicit terminations. However, it warrants pointing that although the approaches to doing so are often highly questionable, the ideas still are explicitly selected or rejected during this phase. According to the model, however, the results of this phase are likely to appear quite as they are expected to appear.
- Development and production phase: The development phase is easily grasped from a methodological perspective. Of the four phases in the funnel model of innovation, the development phase has been analyzed the most and exhibits a correspondingly higher degree of maturity. Naturally, the development may run into any number of difficulties that could force a project termination. For instance, it may not be possible to manage certain technological challenges or one might be faced by budget overruns. Project terminations for such reasons are to be expected and can often be observed in practice. These terminations are almost always a matter of explicit decisions made in the context of project progress meetings. That being said, a closer look at real-world conditions reveals

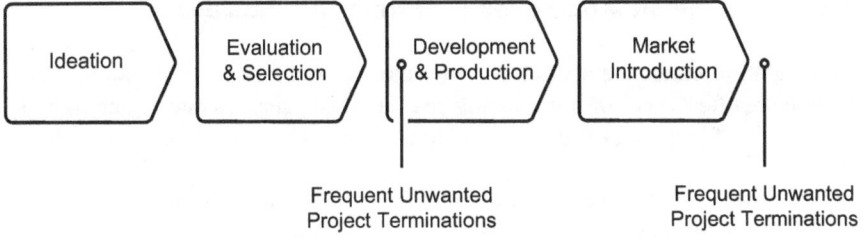

Fig. 3.1 Common points of project termination in the innovation process

the following further phenomenon: many of the ideas selected in the evaluation phase never become the subjects of serious development efforts. Instead, they enter the development phase, but are only halfheartedly pursued within the enterprise and tend to gradually lose relevance as other development projects gain greater attention. From the perspective of company management, such project terminations are essentially unwanted and tend not to be explicit. Moreover, this sort of innovation-project suffocation is extremely common.[1] A closer examination of this point in the innovation process may prove to be highly revealing.

- Market-introduction phase: While serious differences of opinion may arise in the run-up to market introduction, relatively few project terminations occur during this part of the innovation process. Much will have been invested by this point in the innovation process and decision makers will be inclined to push projects through for this reason alone, although we all know that the failure rate for newly introduced products is high.[2]

This short analysis of the various phases in the innovation process shows that unwanted project terminations or failures frequently occur at two points in the innovation process: after selections are made for development and after market introduction (see Fig. 3.1).

▶   Unwanted project terminations are frequent at two points in the innovation process: after selection for development and after market introduction.

---

[1]This seems quite plausible because it is the course of least resistance. For various reasons, it takes courage to officially terminate a project. One prefers on the one hand to avoid demotivating employees. On the other hand, the gradual suffocation of projects may also be a sign of a basic uncertainty or a lack of understanding. One simply does not know how to approach uncomfortable and incalculable projects.

[2]The market introduction phase is characterized by an implicit compulsion to keep going forward, so as to justify and protect substantial investments. One prefers to give it a run and let the market decide. Market failure is easier to accept for the project team and easier to communicate to the executive board.

What Does the Situation Look Like from the Outside?

Looking at an enterprise's situation from the outside, failures after market introduction are often very apparent. In any case, a market success rate of one in five is thought of as favorable. Pessimists tend to estimate the success rate at closer to one in ten.

What then is the success rate after the evaluation and selection phase at the beginning of the development phase? This situation is hardly visible from the outside of enterprises in question. And many people inside the enterprises tend not to want to quantify the matter. Few have an interest in shining a spotlight on these cases. Some will be happy that certain new projects are terminated, and for others it may simply be embarrassing that their projects have failed. In general, such terminations can tarnish the careers of those responsible in the enterprise. They therefore tend to be set aside unceremoniously—and without the numbers—to die a quiet death.

This practice naturally has consequences. Enterprises largely ignore unwanted project terminations at the beginning of the development phase because they are scarcely visible to the responsible managers. When evaluating their own capacity to innovate, enterprises primarily take account of the rate of success for market introductions. The promulgated success rates are therefore significantly better than the real success rates because they do not take account of all of the unwanted project terminations at the beginning of the development phase.

▶   Enterprises largely ignore unwanted project terminations that occur at
       the beginning of the development phase.

This situation leads to yet another significant outcome. The mechanism described for unwanted project terminations only affects those projects that have gone through the evaluation and selection process phase. These normally include projects that come from the research and innovation departments of the enterprises in question. In contrast, incremental innovations of the sort geared to product maintenance are mostly fast-tracked to the development process without first being subject to an explicit, evaluation-and-selection process.

This results in a predicament for the enterprise management. The innovative capacity of the enterprise is overrated while those responsible for innovation remain dissatisfied with the impact their innovation departments have on business development. This also corresponds exactly to the situation that is apparent from the outside.[3]

---

[3]While there are key data that can be used to measure innovative capacity, these tend to be rather indirect (e.g. sales of products that have been on the market for less than 2 years) and their significance is therefore limited. This is why the identification of key data for innovation has remained a subject of research in its own right.

▶   The innovative capacity of enterprises is often overrated while those
    responsible for innovation remain dissatisfied with the impact their
    innovation departments have on business development.

What Does the Literature Say?

What does the current literature have to say about this subject? The two areas we
have concentrated on are treated very differently.

The problems associated with the market introduction of new products are well-
known and have been examined in detail. The same applies to the subject of
remaining successful on the market for the long term. For instance, Geoffrey
Moore offers a very compelling account of the problems associated with market
introduction (Moore 1995) and Clayton Christensen introduces the concept of
disruptive market innovations, describing their surprising properties on the market
(Christensen 1997). It appears that the circumstances that are crucial to successful
market introduction and the long-term success of products on the market are well
understood today. However, it is important not to forget that market introduction
and market success are no trivial subjects and that their implementation also
remains difficult in the real world despite ones understanding of the applicable
principles.

The situation for our second point of concentration, unwanted project
terminations at the beginning of the development phase, looks very different.
There appears to be nothing at all in the literature about this subject. In what
follows, we would first like to examine the reasons for this.

▶   There appears to be nothing in the literature about the subject of
    unwanted project terminations at the beginning of the development
    phase.

Most active enterprise executives are not engaged in the activity of writing of
journal articles and textbooks on the on business topics, let alone topics involving
business theory. New theories are usually drafted and published by professors
working at universities or other research institutions. For their part, professors
tend to lack experience in the actual running of companies. This has led to a
situation in which the latest theories of innovation are based on studies of
innovation activities that university scholars have conducted at enterprises.

Consultants represent another group of authors who have devoted their attention
to innovation. The problem here is that consultants are often systematically unable
to view enterprises from the inside. After all, they too have no experience of being
or having been directly responsible in a managerial capacity within the enterprises
they attempt to support. Their view of the inner workings of enterprises is therefore
similar to that of a university research team.

As we suggested above, there are various internal dynamics that prevent an open assessment and accounting of projects that are more or less left to wither in the early development phase. These same factors naturally make it even more difficult for outside researchers or consultants to access and evaluate unwanted project terminations that occur at the beginning of the development phase. The conspiracy of silence with regard to such project terminations has essentially kept them from the view of outside researchers and consultants and therefore out of the literature. The fact that a significant procedural failure has not been addressed in the literature is therefore not to be attributed to any failings on the part of outside researchers, but to the fact that the phenomenon itself—studiously ignored by insiders—has been opaque to outsiders.

How would it then be at all possible to detect and take account of the mechanism in theoretical contexts? There is only one way of detecting this mechanism. One will have to have had direct experience with the phenomenon in an enterprise. The mechanism becomes visible only when one is personally charged to assume responsibility for innovation management within an enterprise and experience the frustration as a person in charge. It's not something that one shouts out from the rooftops because doing so, as described above, would undermine ones own career. Therefore we now are aware of the phenomenon. But it has yet to be documented. And again, enterprise executives are typically not in the business of writing journal articles and textbooks.

This can only happen when people switch from a career as an enterprise executive to a career in academia. The time spent as an enterprise executive is necessary for an awareness of the circumstances and a subsequent career in academia is necessary to write about the circumstances for purposes of theoretical discourse. Such career changes, however, are rare. And this is why there has been virtually nothing in the literature to take account of the circumstances.

In Daniel Huber, we have one of the exceptions, namely, an individual who has transitioned from innovation management in an enterprise to a career in academia. And this also explains why he is one of the co-authors of this book.

3.2   Interim Conclusion: The Innovation Process Is Incomplete

After an initial rough-and-ready examination of the innovation process, we have discovered that there are two areas in which unwanted project terminations are frequent. The one is well-known and well-documented, namely, the unwanted failure of new products after their introduction to the market (cf. Moore 1995). The literature on this subject is correspondingly extensive.

The other area where unwanted project terminations are common in the innovation process is at the beginning of the development phase, and tends not to be visible from outside the enterprise owing to the reasons outlined above. The area in question is therefore not addressed in the literature. And this, in turn, is why the

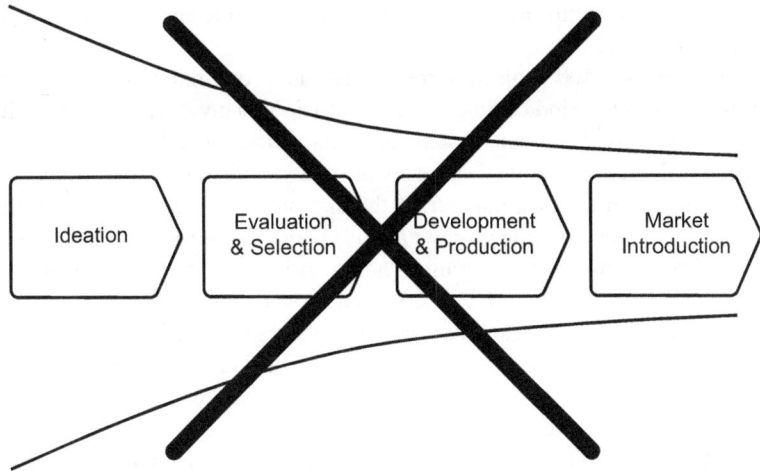

Fig. 3.2 The funnel model of innovation is incomplete

innovation process described in the literature is incomplete and why the process is especially susceptible to unwanted project terminations in practice. This also explains the misgivings one frequently encounters among enterprise executives with respect to their own innovation units.

▶   The innovation process described in the literature is incomplete. This explains why one scarcely observes successful systematic approaches to innovation in real-world enterprise.

Figure 3.2 illustrates the need to expand the funnel model of innovation.

The next chapter offers a description of how unwanted project terminations have manifested themselves after the evaluation and selection phase in the experience of Daniel Huber and the way in which this led to a plausible interpretation of the experience.

3.3   An Innovative Enterprise: Of Explorers and Implementers

As described in the introduction, Daniel Huber worked for more than 20 years in one of the leading innovation centers of the TIME industry,[4] first as an innovator and later as an innovation manager. During this time, he gathered extensive

---

[4]TIME = telecommunications, IT, media and entertainment.

experience with the manifold problems that can undermine and prevent the success of innovation projects.

Between 300 and 400 projects were carried out at the innovation center[5] every year throughout this period of time. On this occasion, many different approaches to innovation were tried out and tested. All innovation-related data were eagerly gathered, applied and examined in terms of their practicability and effectiveness. It is hardly surprising that a number of the approaches proved to be of little use when it came to their practical application. Others, however, proved to be very useful. And an attempt was naturally made to orient the work of the innovation center towards those innovation concepts that proved more reliable in practice. In retrospect, it is clear that these years can also be regarded as a long sequence of "laboratory experiments" in innovation management. That being said, the "laboratory" involved was a part of a real enterprise and the "experiments" were real innovation projects in a real business environment.

Using this long (and unplanned) series of experiments in innovation management as a basis, we would now like to take a look in what follows at the results as these relate to the problem of systematic innovation.

We first discovered that there are actually two ways of looking at the problems associated with innovation efforts. The one way involves an internal view, or a view from within the enterprise, and the other way involves an external view, or a view from the perspective of the market.

The external view encompasses all those difficulties that need to be overcome for a product's successful market introduction. This includes all market-introduction and growth problems, as well as their impact on marketing, strategy, organization and financing. Here, we discovered that the works of Clayton Christensen (cf. 1997) and Geoffrey Moore (cf. 1995) already offer a good explanation for most of the external challenges involved. Their work provided us with very valuable instruments and rules with which to manage our innovation projects.

On the other hand, we detected many complex, internal problems that regularly threatened to block our projects. And there were naturally problems with the fuzzy front end[6] of the innovation process. We anticipated these problems and we discovered many ways of circumnavigating them. A number of these ways are described by Henry Chesbrough in his seminal work *Open Innovation* (Chesbrough 2003, p. 33), while others, such as optimizing the project management and introducing explicit criteria for the evaluation of ideas and idea portfolios, are quite obvious.

On the other hand it was confusing that it proved very difficult to get even the most promising of the selected ideas onto the corporation's agenda, or onto the agenda of the business units in question. Being able to do this, however, is

---

[5]Swisscom Innovations (previously known as Swisscom Research and Development).

[6]Fuzzy frontend refers to the often unclear first two phases of the innovation process (see e.g. Boeddrich 2004 or Kahn 2012, p. 450).

indispensable for the development and successful market introduction of innovations. We naturally anticipated a degree of the not-invented-here mentality, and we indeed experienced this mentality. However, our difficulties went far beyond this. Virtually all of our discussions and meetings proved to be extremely difficult and frustrating. These difficulties even turned up in cases where there were well-developed personal relationships. It was as if the individuals who were familiar with the early phases of innovation projects spoke a different language from that of the remaining employees and managers involved in the decision-making process.

Another problem was that although our innovation center was actually well supported at the highest levels of the corporation, we couldn't avoid the impression that the top managers were not able to really understand us. Even worse, the top managers often seemed to develop expectations with regard to the innovation projects that were clearly unrealistic. It was quite irritating to discover how tenaciously these managers would cling to such unrealistic expectations. And despite direct talks, we were never able correct a single one of these unrealistic views.

As a result, it proved impossible to exploit the potential of the innovation efforts made during the early phases (ideation, evaluation and selection). Many promising innovation proposals were therefore ignored and ultimately lost to competitors. In retrospect, it is clear that this fundamental lack of comprehension is by far the biggest barrier to successful innovation. And here it is important to recall that at least nine of ten innovation projects never reach the market.

▶ The fundamental inability of the innovation center's employees and the corporation's division managers to understand one another proved to be the biggest barrier to successful innovation.

The question now arises as to what exactly occurred in the cases described? What mechanism was at work?

3.4 The Innovation Gap: Why Innovation Projects Systematically Fail

The above-mentioned difficulties were so apparent and systematic that we were forced to confront them head on. In order to be able to discuss the matter at all, we needed to find a suitable way of referring to it. In search of a proper designation for the problem we were experiencing, we were surprised to find nothing in the literature. It then occurred to us that we were dealing with a new problem that had simply not yet been addressed in the literature. We were thereby forced to create our own designation. Originally, we referred to the specific complex of

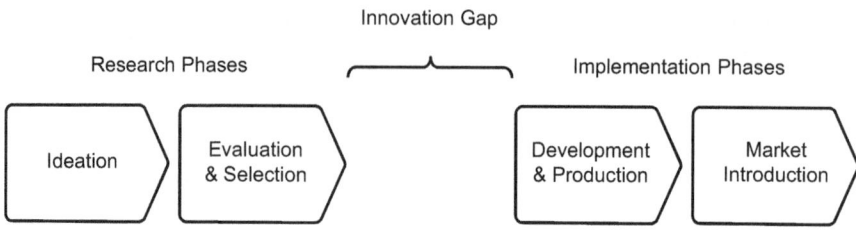

Fig. 3.3 The innovation gap

problems we were dealing with as the *research gap*.[7] This was supposed to express our view that it was a matter of a fundamental gap in understanding and communication that arose between the early research phases of the innovation process (ideation, evaluation and selection) and the implementation phases (development and production, market introduction). Ultimately, however, we settled on the term **innovation gap** to refer to the phenomenon with a more precise term because it is a problem relevant to the innovation process, and actually has little to do with the upstream research activities. Figure 3.3 shows the innovation gap as a gap in the innovation process.

▶   We refer to the gap between the early research phases and the implementation phases as the *innovation gap*.

To meet the difficulties, we decided to implement the obvious measures. We fought the symptoms. We appointed certain individuals with especially well developed communication skills in order to improve the notoriously difficult communication with the top management and the heads of the relevant business units. Similar problems are known in the area of technical sales support, and are handled there by account managers.[8] Following the approach taken in sales, we also appointed account managers in order to facilitate the vexed communication process with corporate executives and the heads of the company's business units. This measure helped—a little...

Thanks to these measures, we were able to successfully maintain our interaction with the management and the heads of the business units. The new account managers were able to detect the difficulties on time and to respond accordingly. Unfortunately, our discussions remained difficult and fraught with

---

[7]We selected an English term because English was the company language. Moreover, a few authors have detected and referred to similar problem in this area. Walter Hehl, for instance, refers to a difficult phase in the same area in German as the "Innovationslücke" (in Hehl and Willmanns 2009, p. 16) and Thomas Hinderling from CSEM speaks of the *valley of death* in connection with startups (Hinderling 2013).

[8]The role of the account manager is a typical role in virtually all sales departments (see Capon 2001).

misunderstanding. Although we had succeeded in professionalizing our approach, the fundamental difficulties remained.

Moreover, new and unexpected problems arose. The researchers and innovators engaged in the early phases were uncomfortable with the role of the new account managers and tended to regard them as unnecessary. Explicit attempts to explain the purpose of the account-manager role also failed! It warrants pointing out in this connection that the innovators belonged among the most intelligent individuals we could find. Added to this was the grumbling from the other side. Many of our extraordinarily intelligent partners in the business units were unable to comprehend why it was so difficult to communicate with the researchers and innovators, even when the account managers moderated.

These perplexing experiences ultimately led us to gain an even deeper understanding of the innovation-gap phenomenon. And this allowed Daniel Huber in the end—indeed, several years after leaving the innovation center—to fully grasp and solve the problem. A certain distance from the fray was necessary.

The Task Structure of Exploratory Work Compared to That of Operational Work

Surprisingly, the first clue to explaining this apparently fundamental communication problem came from an entirely different area of management, namely, from human resources development. An employee at the innovation center experienced surprising and sensational failures in certain projects while achieving excellent results in most projects. This was a source of frustration for the employee. An analysis showed that he was particularly effective whenever his projects had a clearly defined goal. And when they didn't, things seemed to fall apart. In any case, it wasn't possible to pinpoint the exact reasons. Moreover, it turned out that in the further course of the human resources development that nothing could be done to change that pattern. All the same, it proved possible to solve the problem. The employee was only assigned projects whose goals were clearly defined.

What we can learn from this case is that we are confronted by two types of very different tasks:

• Tasks where it is a matter arriving at the best way of reaching a clearly defined goal. This is the typical structure of operational tasks.
• Tasks that consist of the challenge of finding out what a sensible or attractive goal might look like. These are the typical tasks faced by an innovation center.

Figure 3.4 offers a representation of the different tasks.

We also found many individuals who had an aptitude for mastering one of the two different tasks, but usually not both types of tasks. This discovery, however, bore little fruit over the course of many years, i.e. other than allowing us to assign employees more effectively to the tasks at hand.

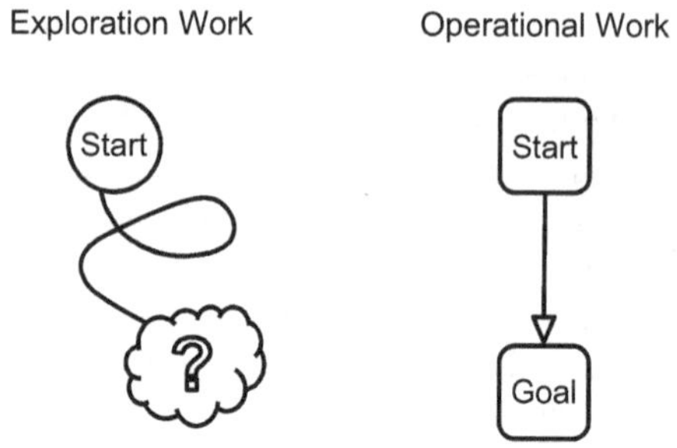

Fig. 3.4 Exploration tasks
and operational tasks

Exploration Work   Operational Work

It wasn't until many years later that it dawned on Daniel Huber why the employees at the innovation center had such difficulties when it came to communicating with employees in the company's business divisions. The two groups apparently had different ways of thinking and approaching tasks. In order to better understand the differences in their thinking, Huber examined the tasks they were typically expected to perform, as well as the processes that they usually deployed to handle their tasks. The comparison of the processes revealed the following fundamental differences:

- Processes for solving operational tasks are typically very elaborate and detailed. They tend to be highly optimized and supported by relatively few, but carefully selected tools and instruments.
- Processes for solving research or exploration tasks tend to be very generic and simple, and they tend to rely on the support of a wide range of different tools and instruments. The reason why this is the case is obvious. If you work at an innovation center and you don't know the result is in advance, then you will not have had an opportunity to develop an elaborate and optimized process for approaching the task, and you will have to make do with a generic process.

Based on the clear differences in the processes, Huber reasoned that there would be corresponding differences in ones assessment of the human qualities of the employees assigned to perform the respective tasks, with certain qualities being judged favorably as virtues or negatively as vices, depending on the type of task at hand.

For instance, it is important for employees who are expected to perform operational tasks to adhere strictly to the specified and optimized processes. Indeed, any departure from the established processes will be seen as posing a risk to higher levels of efficiency and quality. Such tasks require discipline. It is therefore only natural for managers to select certain types of employees to perform such tasks. Discipline and compliance with procedural details will be regarded as virtues when it comes to task performance. A tendency to improvise will be seen as a clear vice.

A different world order applies to employees who are expected to perform research or exploration tasks. Strict adherence to the necessarily generic processes in this context will scarcely be advantageous. Instead, it is important to be able to react flexibly to any new situation that may arise unexpectedly in the course of the

Fig. 3.5 Virtues and vices

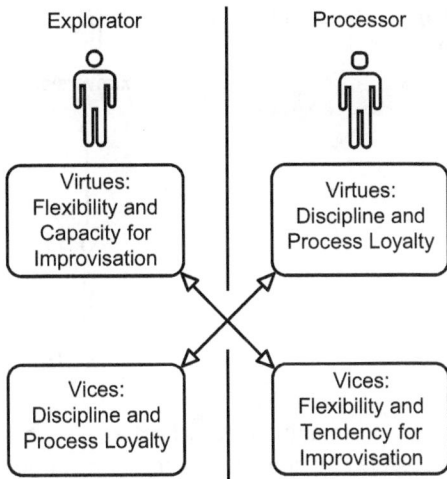

project. Departing from the established processes in this environment is not a problem, and may represent the only possibility of advancing the project. Flexibility, a capacity to improvise and broad-based knowledge are necessary and will therefore be regarded as virtues. A tendency to cling stubbornly to certain templates will be regarded as a vice. And in this case, too, it is only natural to select the correspondingly talented employees to handle the tasks at hand.

Figure 3.5 offers a summary of the findings outlined above. The virtues of an effective explorer are the vices of an effective processor and the virtues of an effective processor are the vices of an effective explorer.

It turns out that business divisions tend to concern themselves with operational tasks and innovation centers tend to concern themselves with exploratory tasks. While this essentially clarifies the issue of why misunderstandings occur so frequently between the two groups of employees, it is however not yet clear why it shouldn't be possible to overcome this regrettable situation. But experience shows that all of the reasonable attempts to do so in practice largely fail.

Typology: What Psychology Tells Us About Personality Types

As described above, employees are selected according to their skill sets and to otherwise create a match for the tasks at hand. However, this also means that very different individuals and skill sets are needed when the nature of the tasks involved is very different.

It warrants having a look at this juncture at what researchers in psychology have discovered about the differences between the various personality types. Indeed, many studies have been carried out in this area of psychology, and these studies have led to the development of multiple personality type schemes. While many of

Fig. 3.6 Myers-Briggs type indicator

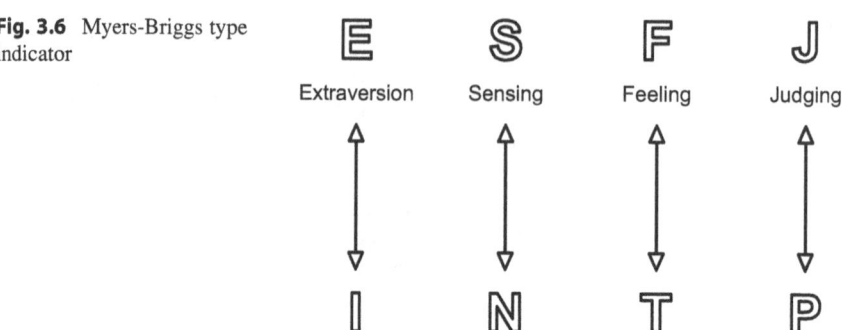

these personality type schemes have been used in the development of highly plausible and coherent explanations of human behavior, it is in our case not really important which scheme to use. To keep things simple, it will suffice for our purposes to limit our discussion to a few of the most common types. We will use in this book the most well known scheme, the Myers-Briggs Type Indicator (MBTI) (Myers-Briggs 1995).[9] The MBTI makes use of four dimensions to break down personalities into various types (as shown in Fig. 3.6).

Myers-Briggs Type Indicator®, MBTI[10]

The Myers-Briggs Type Indicator® (MBTI) is based on the theory of psychological types espoused by C. G. Jung and postulates the existence of four principal psychological functions by which humans experience the world:

- Where an individual's mental focus lies: on the external world (**E** = **extraversion**) or the internal world (**I** = **intraversion**).
- How an individual makes use of information: focusing on the available data per se (**S** = **sensing**) or focusing on their meaning, i.e. on an interpretation of the available data (**N** = **intuition**).
- What an individual uses as a basis for decision making: either on the basis of logical thinking (**T** = **thinking**) or more on the basis of the circumstances and their impact on others (**F** = **feeling**).
- The needs of individuals in accordance with their personal structures: a preference for clear (yes-or-no) decisions (**J** = **judging**) or for leaving all options open for as long as possible (**P** = **perceiving**).

---

[9]See also http://www.myersbriggs.org, as referenced on May 30, 2014.

[10]See   http://www.keirsey.com/handler.aspx?s=keirsey&f=fourtemps&tab=3&c=overview, (as referenced on April 6, 2010).

When cast in their most general form, we can take a binary approach to these four dimensions. Individuals are accordingly either extroverted or intraverted. While real individuals are naturally some blend of the two poles, a rough-and-ready digital model will suffice for our purposes. This simplified personality model will therefore lead to a total of 16 personality types (see footnote 10).

To simplify our account yet further, we can form four groups as follows: SJ, SP, NT and NF, i.e. all individuals with an S and a J in their profiles are assigned to the SJ group.

In his book *Wild Duck* (Dueck 2000), Gunter Dueck describes the impact of the personality types on the management of enterprises. He suggests that we are likely to find only two of these personality groups represented at the executive level of enterprises, namely, the groups SJ and NT, with the SJ type accounting for a clear majority of the enterprise leaders, or around 60%, and the NT type accounting for a clear minority of these leaders, or around 25% (cf. Dueck 2000, p. 100).

Both Gunter Dueck and David Keirsey[11] describe the SJ and NT types consistently in the following manner (paraphrased from Dueck and Keirsey):

- *SJ type: SJs are the major organizers. Keirsey refers to them as "guardians." SJs are convinced that they "do it right". They can be regarded as the pillars of the established system. They are conservative, orderly, reliable and responsible. SJs are analytic and deductive in their thinking. They think in linear chains of cause and effect. They optimize and strive primarily for efficiency. SJs value hierarchical enterprise structures and easily find their stations in the structure. They are risk-averse, and therefore tend to be opposed to change. In short, SJs make for ideal managers and they perform brilliantly when it comes to operational tasks. According to Keirsey, SJs account for around 40–50% of the population.*
- *NT type: In contrast, NTs are the keepers of the key to knowledge. Keirsey refers to them as the "rationals". NTs are convinced that they understand how things work. They can be regarded as the experts. NTs are creative, concept-oriented, professional and impersonal. NTs typically learn without interruption, without ever giving a thought to ever putting what they've learned to use. Knowledge, for them, is everything. NTs typically think in analogies and work inductively. NTs tend to disdain hierarchies and instead seek to form networks. They are likely to experience change as an interesting experiment. In short, NTs make for ideal researchers and perform brilliantly when it comes to unfamiliar terrain. According to Keirsey, NTs account for around 5–10% of the population.*

We therefore see that the differences in the type of tasks described above lead to the selection of SJs for operational tasks and NTs for exploratory tasks (see Fig. 3.7).

---

[11]See http://www.keirsey.com/handler.aspx?s=keirsey&f=fourtemps&tab=3&c=overview (as referenced on April 6, 2010) and http://www.keirsey.com/4temps/overview_temperaments.asp (as referenced on April 16, 2014). The older description used here from Keirsey.com from the year 2010 (http://www.keirsey.com/handler.aspx?s=keirsey&f=fourtemps&tab=3&c=overview) appears to us to be clearer and therefore more appropriate for our present purposes.

Fig. 3.7 NT and SJ
personality types

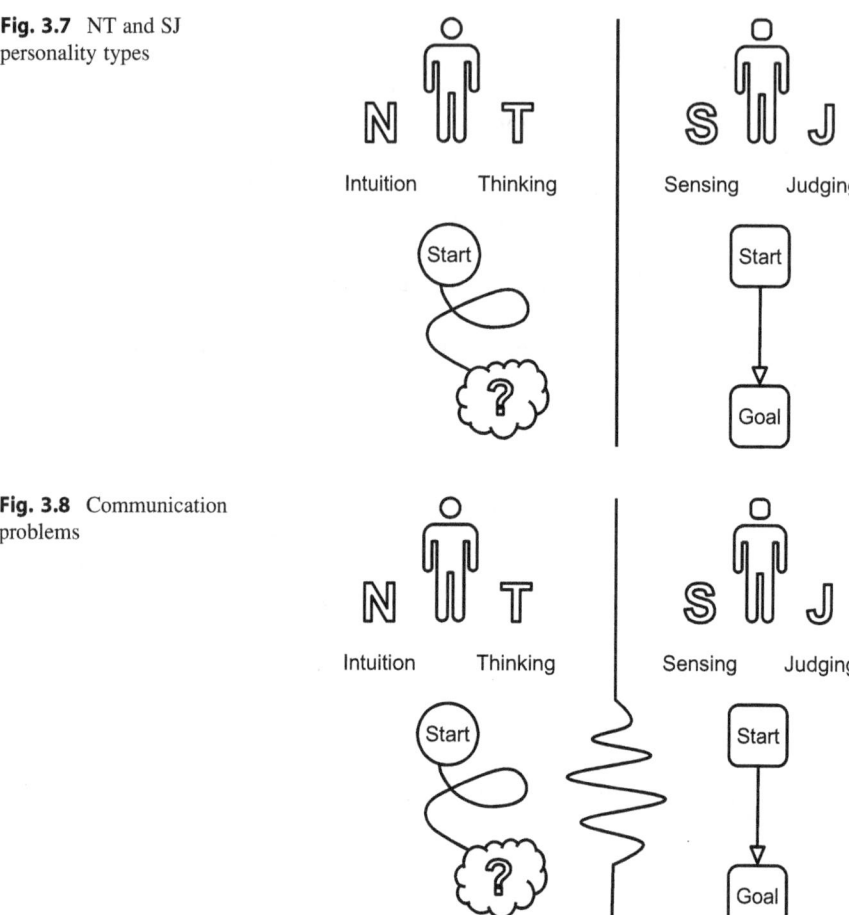

Fig. 3.8 Communication
problems

The selections then shape the enterprise culture that prevails in the respective organizational units. This includes the above-mentioned fundamental differences in the corresponding value systems (i.e. what qualifies as a virtue or vice) as well as profound differences work styles.

Serious communication problems can be expected to arise between SJ-run company units and NT-run units when the two are forced to work together on joint projects (see Fig. 3.8). In his book *Wild Duck*, Gunter Dueck describes the typical misunderstandings that arise between SJ and NT individuals. Dueck's description is entirely consistent with our experience at the innovation center.

▶   SJ types tend to be recruited for operational tasks and NT types for exploratory tasks. One can therefore expect misunderstandings to arise between employees who work in an operational capacity and employees who work in an exploratory capacity.

We can now explain the fundamental communication problem between an enterprise's exploration units (e.g. our innovation center) and its operational units (e.g. business units) and with it the "results" of our long series of "laboratory experiments" in innovation management. We can thereby also explain why such misunderstandings were so extensive on both sides and why the learning curve remained nearly flat despite the extended contact between the groups.

3.5   Conclusion and Proposition 2

We have now identified the source of the problems enterprises experience when facing the challenge of innovation. The classic funnel model of the innovation process that prevails in the literature is incomplete. A failure to account for this incompleteness, and respond accordingly, leads to unwanted project terminations at two specific points in the innovation process, namely, during the early development phase and after market introduction. This explains why we cannot observe virtually any systematic innovation in real-world enterprise settings.

While everyone is aware that new products may very well fail after being introduced to the market, unwanted project terminations during the early development phase remain a largely hidden phenomenon. In our discussions so far, we have been able to explain why this is so. We have adopted the term "innovation gap" to explicitly refer to this heretofore unanalyzed phenomenon. Most innovation projects are terminated during the two above-mentioned points in the innovation process.

It follows that if we want to improve the innovative capacity of enterprises, we will first and foremost need to address these two points in the innovation process: market introduction and the innovation gap. The various difficulties that can arise in the context of market introduction have been discussed extensively by Geoffrey Moore and others in the literature (Moore 1995). In contrast, little has been written about the innovation gap. In our examination of this gap, we have discovered that it arises from a fundamental difference in the personal traits of employees who work in the capacity of researchers and explorers in enterprise innovation units on the one hand, and employees who perform operational tasks in corporate business units on the other. This now leads us to formulate our second proposition:

▶   Proposition 2: There is a heretofore unexamined gap between the early phases of the innovation process known as the ideation and evaluation-and-selection phases in the funnel model and the later implementation phases known as the development-and-production and market-introduction phases.

The question now arises as to the extent these findings can be applied to enterprises interested in improving their innovative capacity. In order to answer this question, we will first need to find out the extent to which the accumulation of NT individuals in exploratory enterprise units and the accumulation of SJ individuals in operational enterprise units represent an organizational necessity or whether they are simply a matter of a misguided organizational solution. We address this issue in the following chapter. We call it the Innovation Gap.

3.6   The Case of Swisskey

The case of Swisskey shows that it is not unusual for path-breaking ideas to remain grounded by uncertainty and development terminations for more than ten years before finally experiencing a breakthrough.

The case begins in 1992, i.e. at a time when the increasing significance of the Internet for electronic business transactions had become evident and attention had been turned to web security concerns. The leading Swiss telecommunications company, Swisscom, launched an internal research project to address this very issue of security. The public-key infrastructure was discussed and people in the company and elsewhere began to grasp the significance of digital identities. The research led to a number of important findings that were then communicated to the executive management. The executive management, however, showed little interest in the research team's findings and proposals.

Convinced that they were onto something important, the members of the research team decided to continue their efforts in the form of a project, although their resources were very limited. They first researched the issue of secure e-mail, including the necessary functions of registration authority and certification authority. This work gradually led to a big picture for their endeavor.

The team's activities were then essentially re-legitimized in 1995 when the executive management issued an order for the development of a smart-card strategy, which necessarily also included the issues surrounding digital identities. When the subject was then submitted a second time for consideration, the executive management recognized its significance. The efforts involved were reinforced and innovative products and market roles to be exploited by Swisscom were sketched.

The project led to the founding of Swisskey in 1998, a joint venture launched by Swisscom and Telekurs (a stock market-data company founded by a group of Swiss banks). A new team was appointed for the company and the results of previous efforts were set aside. After the team had worked autonomously in the relevant area for a time without any clear progress, the Swisskey management concluded that the provision of digital identities did not represent a convincing business case, i.e. it couldn't be turned into a profitable business. Various options were examined and the company's shareholders hoped to win over additional partners. Swisscom was generally unhappy about the influence of the banks in the context of the venture because they had insisted on maintaining very strict security specifications for digital identities, and withdrew from the venture altogether in 2001, effectively ending all Swisskey business activities.

This led to an outcry at many levels in Switzerland. E-commerce was regarded as the latest major development and many feared that the country would miss an opportunity to be a part of the development. Ultimately, the Swiss parliament forced the government to take action.

This attention stimulated new activity. The basis for the legal validation of the digital signature was established in 2003. Corresponding legislation was then enacted and came into effect on January 1, 2005. Many hoped that this move would provide the necessary impetus.

People had at last come to the realization that digital identity required action on the part of the government. In consultation with the Swiss Federal Office of Information Technology, Systems and Telecommunication (FOITT) and the Swiss State Secretariat for Economic Affairs (SECO), progress was made on the exact specifications. Alarmed at declining revenues from its traditional business and encouraged by the vision of an electronic *registered* letter, the Swiss Postal Service helped to get the project back on track. A new public enterprise known as SuisseID was founded to ensure neutrality. Issuance was to be handled by the certifying agencies QuoVadis and Post Digisign. The Postal Service's dense network of branch offices was important for the physical, real authentication of persons desiring a SuisseID. Interested individuals were required to appear in person and present positive identification.

At the time of its launch in 2010, SECO had to fund the issuance of the SuisseID to ensure a distribution of at least 100,000 keys on the market. Owing to its limited applications, the advantages of system use were not obvious and purchases were often made for reasons of prestige, i.e. there was an interest in appearing to be a pioneer. Since 2010, SuisseID has been available on the market. Although it is a good product that is reliable and legally binding, its distribution and acceptance have remained minimal because its use is overly complicated.

How are we to understand all these difficulties? The Swisskey project was confronted by many stop-and-go situations. Multiple shifts in focus were necessary before a usable solution began to take shape. Digital identity was important in the beginning. Then it was primarily a matter of the application. Then the focus shifted back to identity. For a time, technical feasibility was prioritized. And progress was made so long as the technical focus was dominant and the work was a matter of finding the right solutions and means of implementation. Business development, on the other hand, was neglected. And whenever questions were posed as to what the benefits would be for the provider and how the service was supposed to fit in with the provider's strategy, the project showed signs of faltering.

It is also typical that Swisskey was driven in its individual phases by different sponsors and teams: a telecommunications provider with existing customers, a joint venture with a financial services provider, the Swiss Postal Service, the state operating in the capacity of a regulator, a software company and a public association as a neutral superordinate entity. In addition to questions about the right solution, questions about the right promotor also needed to be addressed.

Despite its checkered history, the project nonetheless qualifies as a success story. It depended both on the protective environment of a major corporation in its initial phase and the agility of a startup at a later phase. It took a while to realize that Swisskey could not be operated as an independent company because of its infrastructure or basic-services character. Government representatives needed convincing that it was a new and important service.

One further hindrance centered on the fact that interest in the service was limited. There were few applications that made use of the new function. A critical mass of users was therefore missing. Indeed, it was necessary to develop the market from square one, i.e. beyond those users who incorporated digital identity into their applications more or less for reasons of prestige (the need to appear modern) and not for reasons of broad commercial use.

The government administration itself was also not ready for the use of digital identity after the legal basis had been established, this although it was legally required to accept the digital signature as having the same validity as hand-written signatures. In order to be able to receive digitally signed documents, the administration first had to adapt its document management processes. In particular, it was necessary to configure the electronic repository for the proper administration and archiving of digital signatures. For a long time, the administration regarded this step as an onerous obligation and failed to grasp, especially with respect to the commercial registries, the market opportunities. As the guarantors of the legitimacy and transparency of business transactions, the commercial repositories have an ideal and well-established starting point. There is potential here to expand the business model and to broaden the scope to include other information (e.g. from enterprise signature regulations to spending limits). These new opportunities, however, have so far been left undeveloped.

Conclusion: from the idea and its initial development in 1992, it took until 2010 for a reasonably promising market-ready product to emerge. There were many project terminations and restarts, including the founding of companies and their subsequent liquidation. Previously acquired know-how was only weakly exploited during the project resuscitation phases. All of these factors are very typical of real-world innovation activities, and figure in the histories of many successful innovations.

Our use of the Swisskey case is meant to illuminate a number of important aspects of innovation: perseverance pays; all potentially disruptive projects go through official and clandestine phases; the common rationale for terminating a project "we already tried, and it didn't work" is usually groundless. The circumstances that an attempt was made and yet the subject has remained tantalizing is more of a confirmation of an idea's potential, although no one has yet found the key. Problems may also arise when projects conflict with the latest dominant business model, or when they call for something entirely new that is not yet sufficiently understood.

References

Boeddrich, H.-J. (2004). Ideas in the workplace: A new approach towards organizing the fuzzy front end of the innovation process. *Creativity and Innovation Management, 13*(4), 274–285.

Capon, N. (2001). *Key account management and planning*. New York: Free Press.

Chesbrough, H. (2003). *Open innovation*. Boston: Harvard Business School Press.

Christensen, C. M. (1997). *The innovator's Dilemma*. New York: Harper Business.

Dueck, G. (2000). *Wild Duck*. Berlin: Springer.

Hehl, W., & Willmanns, R. (2009). *Paradoxa und Praxis im Innovationsmanagement*. Munich: Hanser.

Hinderling. (2013). http://www.kmu.admin.ch/aktuell/00524/00730/01247/index.html?lang=de. As referenced on May 11, 2013.

Kahn, K. B. (2012). *The PDMA handbook of new product development*. Hoboken: Wiley.

Moore, G. A. (1995). *Inside the Tornado*. New York: Harper Business.

Myers-Briggs, I. (1995). *Gifts differing: Understanding personality type*. Palo Alto: Davies-Black Publishing.

"Good Management" in Enterprises Today May Be Blocking Innovation

4

4.1 Corporate Culture and Values Such as Championed in the Literature

Enterprise culture (also referred to as corporate or organizational culture) is often described as a decisive factor for successful innovation in the relevant literature (e.g. in Vahs and Brem 2013, p. 190 ff.; Müller and Dörr 2011, p. 17 ff.; Zillner and Krusche 2012, p. 251 ff.). Edgar Schein qualifies as the person who coined the term corporate culture. Schein defines corporate culture as follows:

- *A pattern of basic assumptions—invented, discovered, or developed by a given group as it learns to cope with its problems of external adaptation and internal integration—that has worked well enough to be considered valid and, therefore, to be taught to new members as the correct way to perceive, think, and feel in relation to those problems (Schein 1985, p. 9).*

Further statements on the subject of enterprise culture include:

- *The culture concept has been borrowed from anthropology, where there is no consensus on its meaning. It should be no surprise that there is also a variety in its application to organization studies (Smircich 1983, pp. 339–358).*
- *Organizational culture is the collection of traditions, values, policies, beliefs and attitudes that constitute a pervasive context for everything we do and think in an organization (Marshall and McLean 1985, pp. 2–20).*
- *This is how we do things around here (Bright and Parkin 1997, p. 13).*

© Springer International Publishing AG 2017

D. Huber et al., *Bridging the Innovation Gap*, Management for Professionals, DOI 10.1007/978-3-319-55498-3_4

In any case, our aim is to write a book about innovation and not about enterprise culture. We therefore limit our discussion of the term enterprise culture to the extent necessary for an understanding of the innovation process.

Still, it is safe to assume that enterprise culture will have a significant impact on the innovative capacity of enterprises, although perhaps in a somewhat more complicated manner than has been thought. Vahs and Brem (2013, p. 190), for instance, present many examples of the ways in which enterprises describe their own organizational culture and their stance with respect to innovation. The content of these statements can essentially be paraphrased as follows: "Innovation is an important part of our enterprise culture." Such high-minded, self-defining statements of the sort one finds in the official profiles of virtually every enterprise can be useful in the context of company management and development. However, it is also quite clear that enterprise cultures cannot simply be decreed from above. They need to grow from below, as Edgar Schein's definition of organizational culture suggests (see also Zillner and Krusche 2012, p. 251 ff.). While such guiding principles indeed say something of the goals of the enterprise management, they may not be especially reflective of an enterprise's existing culture.

▶   Enterprise cultures cannot simply be decreed from above. They need to grow up from below.

While the aim of most publications on the subject of enterprise culture and innovation has primarily been to explain how to establish an innovative enterprise culture, we conclude the following in Chap. 3: "*SJ personality types tend to be recruited for operational tasks and NT personality types for exploratory tasks.*" This means that not only one but two ways of thinking (both SJ and NT) are necessary for innovative enterprises. Given this somewhat surprising result, we have to ask ourselves, as we did at the end of Chap. 3, about the: "*extent to which the accumulation of NT individuals in exploratory enterprise units and the accumulation of SJ individuals in operational enterprise units represents an organizational necessity, or whether it is simply a matter of a misguided organizational solution.*" We now turn our attention to answering this question.

▶   Innovative enterprises need two ways of thinking, not just one: SJ thinking and NT thinking.

4.2   Corporate Culture in Real Enterprises

As we saw in the last chapter, SJ types make for ideal managers of operational business and NT types are especially well-suited to performing exploratory tasks. This allows us to answer the question we posed above. It appears to be an organizational necessity because enterprises require both operational excellence and a capacity for discovering new fields of activity. This means that the differences

Fig. 4.1 Cultural conflicts

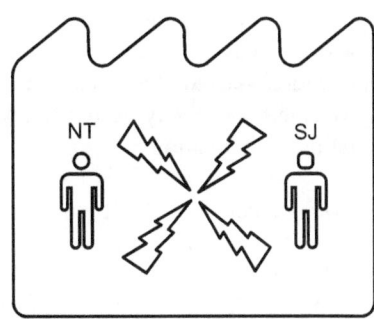

in corporate culture described above must be a necessary property of any innovative enterprise! Exploration and research don't work without NT types and successful operational business doesn't work without SJ types.

We are thereby faced by the surprising requirement of having to simultaneously implement two different enterprise cultures in the same enterprise: an NT culture for research and exploration and an SJ culture for the rest of the enterprise, i.e. for the conduct of operational business. This requirement, however, contradicts the precepts of "good management" both with respect to the organization of enterprises and brand maintenance. It is said that you can't operate an enterprise on the basis of two different regulation and value systems. And you can't project two different value systems on the market.

▶   Two different corporate cultures need to be implemented simultaneously to establish or maintain an enterprise's innovative capacity.

Figure 4.1 shows the resulting cultural contradictions and conflicts.

How can we avoid these conflicts? Let's first have a look at what happens when we don't intervene with any specific measures:

- We observe the conflict between the two cultures.
- The stronger of the two culture gains the upper hand.
- The stronger of the two cultures is the one that is supported by more employees and that appears to bring more money into the enterprise. This dominant culture is the SJ culture in both cases.
- The development weakens the bond that the NTs have to the enterprise. They feel misplaced and begin to leave the enterprise.
- This undermines the enterprise's capacity for innovation because research and exploration are no longer pursued in the necessary quality and quantity.
- Certain NTs conform. There remains a pool of employees who are more or less capable of speaking both languages. It is precisely these employees who allow the enterprise to display a minimal capacity to innovate.
- This makes the innovative capacity of the enterprise dependent on these few key individuals.

These developments can be expected when the system is left to itself. This however looks very familiar to those whose work centers on the management of innovation. Indeed, the scenario described above corresponds exactly to the state of most enterprises today. And when we take a closer look, we find the following facts relating to innovation:

- Many enterprises don't innovate at all.
- Many enterprises do a very poor job at innovating.
- Many enterprises innovate by default in an ad hoc manner. Their innovation is typically depends on the engagement of a few individuals (often the enterprise owners).
- Many enterprises have (often excellent) research units that are poorly linked to the remainder of the enterprise.
- Many enterprises simply endure the internal conflicts between SJs and NTs.

We see that we essentially wind up with what's out there today when we don't do anything. There is naturally nothing surprising about this finding. After all, nothing has yet been undertaken in real-world enterprises to bridge the innovation gap.

4.3   Conclusion and Proposition 3

We have so far seen the following:

- Enterprises are not directly aware of the innovation gap because they are unable to perceive it. They don't experience it as an independent transition phase, but are simply relieved when they reach the other side where things look familiar again.
- Enterprises are not prepared to manage tensions between the forces of NT and the forces of SJ.
- This leads to a cultural conflict that ends in SJ dominance.
- This undermines the enterprise's capacity for innovation.

Upon closer examination of the mechanisms involved, we find that if businesses were to be run according to ideal laws of business management, they would be run exclusively according to SJ principles. While such enterprises would be highly efficient, they would be nearly incapable of innovation! This conclusion leads us to our third proposition:

▶   Proposition 3: Enterprises managed according to conventional principles of business management are likely to be incapable of innovation!

4.4   The Case of Bluewin

Swisscom's internet service provider (ISP) Bluewin offers an excellent example of how innovation is hindered when the cultural conflict between SJs and NTs cannot be resolved, and how things can take a turn for the better despite unfavorable conditions.

The case begins in 1995 at a time when the telecommunications market was undergoing major changes that also promised to richly reward those companies that managed to keep abreast of developments. There was a heightened sense of needing to exploit the opportunities associated with the new technologies. The major players were on the lookout for market opportunities in new business fields to compensate for reduced profit margins in traditional business segments. The appearance on the scene of young, innovative companies was met with suspicion and there was considerable anxiety about missing out on opportunities. This point in time also marks the beginning of those activities that culminated in the range of Internet services offered by Swisscom today.

Things got rolling when an executive at Swisscom (at the time still known as PTT) joined an entrepreneur to set up what was then a novel content-oriented Internet portal. With Swisscom's approval, this took place explicitly outside the corporation because there was concern that the corporation's internal decision-making procedures might prove to be too sluggish for this time-is-of-the-essence venture. The vision was to create a content portal that was to be financed on the basis of a subscription business model. The first step in the start-up phase was therefore to appoint an editorial team to develop the content for the portal. Although the corporation was aware that something significant was in the making, it was uncertain as to exactly how it should approach the matter.

The project was soon rated as important enough to insist on its integration in the corporation. The portal was introduced to the market under the brand name Blue Window. A decision was made, however, to delay making access to the portal contingent on the purchase of a subscription. The service would therefore initially generate no revenue. This was something that was completely out of sync with the corporation's standard business model. It was therefore regarded as a bit of an exotic bird and inspired considerable ambivalence on the part of the management.

Once it became ever clearer that the Internet was set for broad-based growth, a decision was made to invest robustly in Blue Window. Quite in keeping with an NT game plan, the developers experimented with communities and gathered experience in the burgeoning segment of internet-based media and virtual electronic shops. All of this activity was something completely new and unusual for the corporation. And the lack of a shared understanding of what was at stake was also a glaring fact. On one side, there was NT culture in support of Blue Window, and on the other side was SJ culture fretting about a corporate sell-out. And all this with still no functioning business model in sight (what is not really a problem for pronounced NT types...). These differences and the lack of revenue led to a crisis. On the one hand, there was a confident sense of being on the right path to the future, while on the other hand the pressure kept growing to demonstrate profitability. And many decision makers wanted to put limits on what might turn out to be a foolish venture. Then it came to a change in leadership.

Operating in accordance with traditional enterprise specifications, the new management focused on the establishment of a sustainable revenue model (quite in keeping with SJ principles) and found a solution in a pay-for-access approach. Customers were to be given access in exchange for a monthly subscription fee. While this no longer had much to do with the desired content portal, it promised to generate revenue. And even better, it was a perfect match for the traditional Swisscom business model of network access fees. This move to covering expenses put an end to the difficulties. The editorial team was streamlined, but not scrapped altogether and the new business unit was integrated into the corporation, first under the Bluewin brand and later under the Swisscom brand. Today, Swisscom/Bluewin is far and away the largest Internet service provider in Switzerland and qualifies by all means as a success.

How are we to understand the difficulties? The cultural differences between the NTs and SJs in the present case study are very apparent. It was the NTs who initiated the Blue Window project. And, in the beginning, the new company was dominated by an NT culture. It was the same NT culture that enabled the acquirement of an important knowledge base in the area of Internet-based services, at first without a viable business model. It was then the SJ-dominated culture that paved the way for the consolidation and growth of the business. The new SJ-dominated management also had the wherewithal to retain and to make use of what had been accomplished by the preceding NT culture. One can also clearly discern how the interaction of the two corporate cultures led to misunderstandings during the early phase of development. For those like Daniel Huber who were directly involved, however, the time-consuming and essentially exasperating discussions between the NT employees and their SJ colleagues left a profound and lasting impression.

Conclusion: in retrospect, we can conclude that the SJ-dominated enterprise would not, under normal circumstances, have been able to detect and develop the new business field. The original decision to initiate the venture outside the existing corporate structure proved to be on the mark. On the other hand, we can also confirm that the successful reintegration was only made possible by switching to a traditional business model. The pay-for-access model was actually a stroke of fortune that saved the venture for Swisscom. It is unlikely that it would otherwise have been possible to develop the new business field.

References

Bright, D., & Parkin, B. (1997). *Human resource management, concepts and practices*. Houghton-le-Spring: Business Education Publishers Ltd.

Marshall, J., & McLean, A. (1985). Exploring organization culture as a route to organizational change. In V. Hammond (Ed.), *Current research in management* (pp. 2–20). London: Francis Pinter.

Müller, T., & Dörr, N. (2011). *Innovationsmanagement*. Munich: Hanser.

Schein, E. H. (1985). *Organizational culture and leadership*. San Francisco: Jossey-Bass.

Smircich, L. (1983). Concepts of culture and organizational analysis. *Administrative Science Quarterly, 28*, 339–358.

Vahs, D., & Brem, A. (2013). *Innovationsmanagement*. Stuttgart: Schäffer-Poeschel Verlag.

Zillner, S., & Krusche, B. (2012). *Systemisches Innovationsmanagement*. Stuttgart: Schäffer-Poeschel Verlag.

Part II
The Solution

The Organizational Structure: The Innovative Enterprise

5

In the previous chapter, we arrived at a surprising third proposition, namely, that ideally managed enterprises will actually be incapable of innovation. Is this proposition at all plausible?

Many executives and enterprise observers will resist this suggestion and hasten to point out that innovations are indeed developed in enterprises with a well-structured and professional management.

The aim of our third proposition is naturally not to present a description of the way real-world enterprises operate. It simply shows what one could expect of an enterprise in a simplified, ideal state of management. And, theoretically at least, it makes a lot of sense. For instance, an ideally managed enterprise would also be ideally efficient. Efficiency means in this case that everything that needs to be done is optimized and that nothing is done that is not absolutely necessary. There simply are no unplanned activities in the ideally efficient enterprise, nor is there any time for "idle" reflection and free association. The problem with this is that reflection, free association and spontaneity are a necessary condition for the kind of creativity that is the source of innovation. This reason alone suffices to validate our third proposition.

However, as shown in the previous chapter, the fact that enterprises rely on two different enterprise cultures to enable innovation offers yet another reason why ideally managed enterprises are incapable of innovation. According to the rules of the Art, it simply wouldn't be possible.

We now turn our attention to finding a solution that will enable enterprises to do a better job of innovating despite the obstacles.

All illustrations are published with the kind permission of © Heiner Kaufmann, Daniel Huber, and Martin Steinmann. All Rights Reserved.

© Springer International Publishing AG 2017
D. Huber et al., *Bridging the Innovation Gap*, Management for Professionals,
DOI 10.1007/978-3-319-55498-3_5

5.1   The Organizational Consequences of Systematic Innovation

A Closer Examination of SJ and NT Typology and Culture

In the previous chapters, we learned that certain personality types are especially well-suited to the performance of innovation tasks and certain other personality types are especially well-suited to the performance of operational tasks. In particular, innovation tasks require NT personality types and operational tasks require SJ personality types. We also learned that we need two different enterprise cultures to accommodate both of these types, this although the existence of two different corporate cultures contravenes generally accepted tenets of management theory.

The question now arises as to the consequences of these findings and how we can best approach the situation. Given that there are very good reasons for giving safe harbor to both cultures and personality types, and given that we know from organizational theory that enterprises can essentially provide safe harbor to only one culture, we have no other choice but to establish two organizations, one responsible for innovation where NT culture is allowed to prevail and one responsible for operational business activities where SJ culture is allowed to prevail. This means that we will have to split our enterprise in two parts that are each sufficiently autonomous to maintain their own corporate culture. More specifically, we will need to separate our research and exploration unit from the rest of the enterprise and furnish it with the autonomy it will need to preserve its own system of values! Moreover, the scope of this autonomy will have to include independent business processes, criteria of promotion, incentives, recruitment policies and salary decisions. Figure 5.1 shows the separation of the innovative enterprise unit with its NT-dominant culture from the rest of the enterprise with its SJ-dominant culture.

Fig. 5.1 An autonomous innovation unit

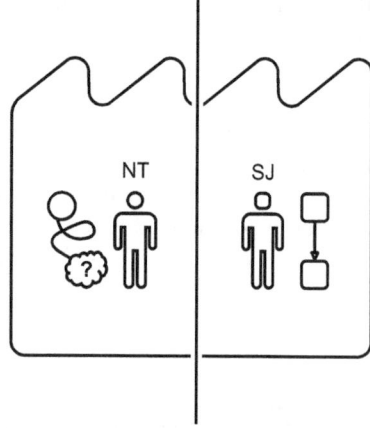

▶   The innovative unit of an enterprise should be separate from and
    autonomous of the remaining SJ-dominant units. Separation and auton-
    omy should enable the innovation unit to maintain its own NT enterprise
    culture with its own system of values.

We have thereby reached two of our goals:

• Both cultures are now free to develop.
• There is no longer a continuous conflict between the two cultures.

This desired result, however, comes at the cost of a new problem: our two
autonomous units now have essentially nothing to do with one another.[1]

We now need to overcome this unsatisfactory state while at the same time
ensuring that both of the cultures that are necessary for success will continue to
be allowed to develop freely. This means that we will have to retain the divided
organizational structure. On the other hand, it will also be necessary to implement
measures that will enable these two autonomous enterprise units to optimally
interact with one another. For this purpose, we introduce an explicit communication
and coordination role. While we are therefore once again confronted by our old
communication problem, the new arrangement is now open and explicit. And this
represents progress because the fact that our communication problem is now in the
open means that we can address it directly. In Fig. 5.2, the existing communication
problems are represented by a waved-shaped line of separation between the two
enterprise units.

So what do we have to do? As we have seen, we now have two enterprise units
that do not speak the same language, as it were. Just as in the case of languages,
what we need is an interpreter. Indeed, we know this situation from our sales
department. This is because customers, too, speak a fundamentally different language. And here, too, we need an interpreter. The literature offers us many proven
approaches to solving this problem, for instance, via key account management
(Capon 2001). What we need to do therefore is to introduce a key account
management to operate between the research and exploration unit and the operational unit of the enterprise.

The question arises as to the side on which the translation is to take place. When
we analyze the thinking of the respective NT and SJ personality types, we find that

---

[1]We have now achieved the state of all those enterprises that permit themselves the luxury of an
autonomous research center. Our discussion so far explains why such research centers often
deliver excellent results, i.e. because they are places where an NT culture is free to develop. On
the other hand, we also have an explanation for the difficulties that the remainder of the enterprises
have when it comes to the commercial exploitation of these excellent results, i.e. on account of the
known communication problem between the two cultures. There are naturally further significant
reasons that prevent the enterprise from commercially exploiting the results of exploration. One of
the most important reasons is that it is often difficult to bring innovation projects into alignment
with existing business models. This is the problem of "white spaces" that Mark Johnson describes
in his important book *Seizing the White Space* (cf. Johnson 2010).

Fig. 5.2 A vexing
communication problem

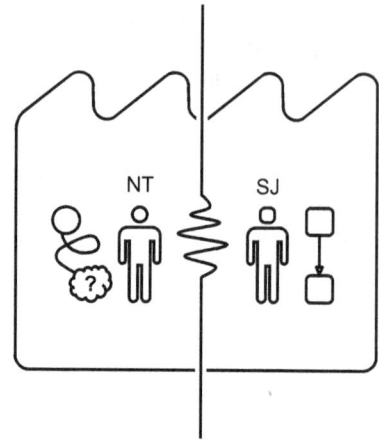

the NT types are quite capable, although perhaps often unwilling, of making use of the linear and highly rational forms of discourse that are characteristic of SJ types. In contrast, the multipolar and intuitive forms of discourse favored by NTs are scarcely conceivable and even alienating for SJ types. This makes it extremely difficult, if not impossible, for SJs to speak the NT language, whereas it is merely tedious or cumbersome for NTs to speak the SJ language. For this reason, we are well-advised to position the role of the interpreter, our key account management, on the side of the NT culture, our research and exploration unit.

▶   Working in the capacity of a cultural interpreter, a key account manage-
     ment is to be responsible for joining the research and exploration part of
     the enterprise to the operational part. This management is to be
     integrated into the NT-dominated innovation part.

This implies that the research and exploration part of the enterprise needs to be responsible for its own marketing. In other words, it will need an independent and visible brand, as well as the accompanying status and the corresponding attention of the CEO and the supervisory board.

▶   The research and exploration part of the enterprise must be responsible
     for its own marketing and communication, i.e. it must have a visible
     brand of its own within the enterprise.

Figure 5.3 summarizes our findings so far. The innovation unit is to be outfitted with a key account management and be responsible for its own marketing and communication.

Fig. 5.3 Key account
management, marketing and
communications

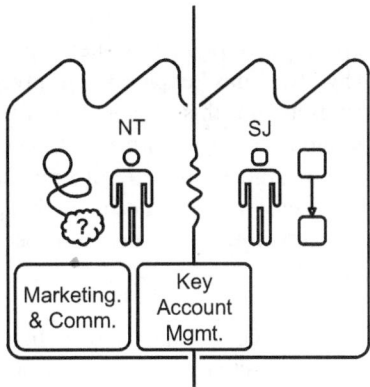

Consequences for the Enterprise Structure

Our analyses have so far yielded results that will have an impact on the enterprise's general organizational structure. We can itemize the impact for the enterprise's organizational structure as follows:

1. The units responsible for research and exploration must be separate and autonomous of the remaining business units if the enterprise is to remain innovative in the medium term. This organizational unit is to encompass all of those parts of the exploration process that precede the development phase. This is the only way of ensuring that both of the indispensable enterprise cultures can develop freely.
2. This separate innovation unit must be entirely independent of the operational units. The scope of this autonomy includes processes, criteria of promotion, incentives programs, recruitment policies and salary decisions! This autonomy is not simply a privilege, but an absolute requirement for the proper functioning of the entire system.
3. This separate innovation unit must be linked to the rest of the enterprise via a key account management that functions as a cultural interpreter. This interpreter function must be integrated in the NT-dominated innovation unit. Here, it makes sense to request a global service order from the executive management for early-warning and exploration projects.

   This order should remain as constant as possible. In other words, it should not be coupled to operational results. In particular, reductions in order volume are to be avoided at all costs.[2]
4. In contrast, the development and market-introduction phases are to remain in the SJ-dominated operational units of the enterprise.

---

[2]Reductions in the volume of orders for early-warning and exploration activities are so damaging because they usually go hand in hand with workforce downsizing. Such staff cuts lastingly undermine the motivation of NT employees, which can lead to a serious decline in an enterprise's capacity for innovation.

This organizational structure is counterintuitive from the perspective of the typically SJ-oriented business management. SJs will find it difficult to understand why "frivolous" explorers are given such freedoms at the cost of efficiency. It is precisely for this reason that functioning innovation structures are inherently unstable! They require an intellectual understanding of the overall system and their legitimacy is therefore regularly called into question within the enterprise. Their raison d'être needs to be reasserted again and again. That is why the description of these specific aspects of the overall system is emphasized in this context.

▶   Functioning innovation structures are inherently unstable. They require an intellectual understanding of the overall systems and their legitimacy is therefore regularly called into question within the enterprise. Their raison d'être needs to be reasserted again and again.

A look at the real-world performance of enterprises confirms these findings. Enterprises often go through phases of heightened innovation and then fall back into extended phases of mediocrity. The reason in most cases is that the liberties enjoyed by innovation units after certain periods of success are curtailed, ultimately resulting in the stagnation described above. Owing to their success, their special freedoms become clearly visible, which, according to SJ logic, naturally calls for immediate correction. It warrants pointing out that the relationship of cause and effect in such turns of fortune is not readily apparent to the managers. After all, for a certain amount of time, enterprises may well be capable of continued innovation based on ideas generated during the periods in which innovative activities received more robust support. Then the point is reached beyond which there is nothing new and inspirational. This pattern occurs because well-functioning innovation groups tend to produce more innovative business ideas than an enterprise can absorb (Chesbrough 2003, p. 33). A backlog of ideas and unrealized project proposals develops. When the cost-saving measures are introduced, the loss remains invisible for a period of time. Managers remain unaware of the medium-term opportunities that have been sacrificed.

Conclusion

Outfitted with an intellectual grasp of the relationships involved, executive managers are called upon to structure their enterprises according to the above-mentioned rules. And they are to do this despite the fact that it is counterintuitive for the members of the management. For this reason, it is also important to make the corresponding decisions in an explicit manner and to comprehensively document the decisions along with their rationale.

▶   It is important to make the decision to restructure the enterprise for
    greater innovative capacity in an explicit manner and to comprehen-
    sively document the decision along with its rationale.

We now know how to structure the enterprise in the interest of promoting
innovation. We have also determined that no new requirements need to be met by
the operational units of the enterprise to enable innovation. In contrast, the exact
structure of and the requirements that are to be met by the innovation unit is in need
of clarification. We therefore turn our attention now to the innovation unit of the
enterprise.

5.2   A Closer Examination of the Classic Innovation Process

If we want to examine the design of the innovation unit, we will first have to have to
concern ourselves with the activities this unit will be expected to perform. In other
words, we will again need to consider the innovation process. As outlined in
Chap. 2, the innovation process is usually described in the literature with reference
to the funnel model. According to the basic model, the process includes the
following four phases: ideation, evaluation and selection, development and production, and market introduction. The funnel model of innovation is a very widely used
model. While it is a very simple and intuitive model, it comes with one major
drawback. It doesn't do a good job at describing our real-world experience! Sadly,
the real world is a bit of a ruffian in this specific case because it systematically
refuses to obey our so very perfect model.[3]

▶   The funnel model of innovation is a very widely used model. While it is a
    very simple and intuitive model, it comes with one major drawback. It
    doesn't do a good job at describing our real-world experience!

In order to underscore this very important finding, we reintroduce the above
figure. The funnel model is incomplete and doesn't correspond to the real world—a
major shortcoming (see Fig. 5.4).

---

[3]We addressed this topic for the first time in our book *Innovation Factory* (Mock et al. 2013). We
review the topic at the present juncture to secure a cohesive presentation of our argumentation.

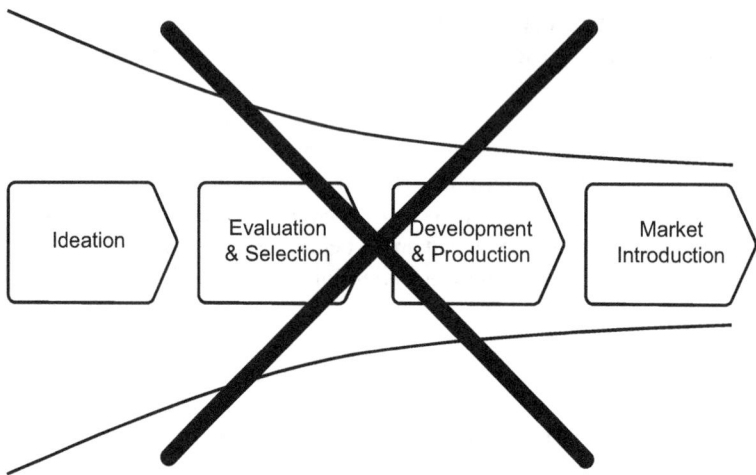

Fig. 5.4 The lack of correspondence between the funnel model and the real world

The Classic Funnel Model of the Innovation Process

Let's first have a look at the theory behind the funnel model:

- Ideation phase: The purpose of the ideation phase is to generate ideas for innovations. Here, we usually apply one of the many known methods of ideation (e.g. brainstorming). In most cases, this is initially a matter of generating as many ideas as possible, even far-fetched ideas.
- Evaluation-and-selection phase: The newly generated ideas are compiled and then evaluated according to their overall attractiveness. Many authors have recommended the use in this context of cost-benefit analyses (see e.g. Eversheim 2003, p. 93). This evaluation leads to a ranking of all of the proposed ideas. Often, the highest ranked idea or ideas are selected and directly forwarded for further processing to the development phase. All of the ideas that have not been selected are either discarded or (better) filed for possible future use.
- Development phase: It is now the task of those in charge of the development phase to turn the idea into a real product. A functional specifications document is usually drafted to accompany the actual technical development. Certain aspects of such prospective products may remain problematic. And there may also be doubts as to whether it will be at all possible to solve the problems. The response in such cases is to carry out feasibility studies, which often require extensive laboratory testing. If the product is determined to be feasible, the next step is to demonstrate this by developing a prototype. The subsequent phases in the product's development include steps to achieve industrial production and any necessary preliminary work and adaptations for production. All of this usually involves a considerable amount of time and financial resources. This is why enterprises are able to afford no more than just a few simultaneous development projects, in most cases only one. The decisions as to which products are to be

developed are therefore of great significance to the enterprise and also harbor considerable risks. These decisions are typically made by the management in a formal context, or even at the level of the supervisory board.

- Market-introduction phase: After feasibility has been established and it is determined that the product can be manufactured, it is time to prepare for the product's market introduction. Depending on the industry involved, this step in the process may also be extraordinarily time and resource consuming. Given that the product will be made visible to the customer upon its market introduction, insufficiently refined products[4] or market communication glitches can undermine the enterprise's reputation and have a lastingly damaging effect on the brand. During the market introduction phase, it is essential therefore not only to make sure that the product can be sold, but also that doing so will not damage the reputation of the enterprise.[5]

As a result of the major investments that are required in the development and market-introduction phases, failures during these phases can have serious consequences for the enterprise. In many cases, failures may even threaten the enterprise's survival. This is why it used to be common practice for enterprises to place the focus on the development and market-introduction phases. As a result, these phases are fairly well understood today as the third and fourth phases in the innovation process. This cannot be said about the two early phases *ideation* and *evaluation and selection*, which received far less attention in the past. Part of the problem was the actual content of these phases, especially their creative aspects, which were difficult to grasp from a process-based perspective. This led, for instance, to the common use of terms such as *fuzzy front end* to refer to these phases.

Criticism of the Classic Funnel Model

As mentioned above, the funnel model of innovation has the disadvantage of not being applicable to the real world.

It may come as no surprise therefore that important distinctions between theory and practice appear particularly in the context of the first two process phases. Let's first have a look at the more important of the two phases:

(a) **Evaluation-and-selection phase**

Those involved in the evaluation-and-selection phase are responsible for determining which of the many possible innovation ideas are to be forwarded to the (expensive!) development phase. As already mentioned, this decision is very

---

[4]The term "product" in our present discussion essentially refers to anything that an enterprise winds up selling, and naturally includes services of all kinds.

[5]The current practice of ceremoniously announcing the imminent arrival on the market of products can lead to overblown expectations and unnecessary pressure that could undermine the resilience of a product still in need of nurturing.

important for the enterprise. After all, the very survival of the enterprise may depend on the successful development of the selected ideas. In light of this importance, the classic model provides for the use of a cost-benefit analysis in the context of idea evaluation and selection. It follows that the most attractive of the many generated ideas is to be identified and selected. This appears at first glance to be a highly reasonable approach. However, the selection does not yet say anything definitive about the commercial value of the idea.

Our experience shows that scarcely any of the generated ideas is really commercially attractive in and of itself. From the perspective of the management, even the very best idea usually comes across as an insignificant detail. This shows that the evaluation of the idea, as provided for in the classic funnel model of innovation, is essentially inadequate.

That being said, the real world of enterprises is a place where valuable innovations do take place. Let's have a look at how these are typically structured in terms of their innovative elements. Here, we find that successful new products are to a large degree based on existing functions. These usually appear in a new arrangement and are supplemented by novel functions. The new product that appears on the market is therefore mostly a system that combines (many) existing and (few) truly novel functions to form a new multifunctional product. It is this new composite system that may generate value for the enterprise.

Figure 5.5 offers an illustration of a typical innovation. The graph includes a bar diagram with a value scale. The first bar represents the commercial value of a single new idea. The value is minimal. The second bar shows the significantly higher value of the idea after it has been integrated into a composite system.

While the new individual function represents only a small share of the value, it is this (minor) new function that is accounted for by our selected idea. If we evaluate only the new individual function, then we may very well be inclined to see a rather modest detail. Without an appreciation of the new composite system, we will not be in a position to estimate the commercial value of the idea. Unfortunately, this composite or overall system is not yet known in the evaluation phase. It is precisely on account of this essential difficulty that the evaluation and selection phase in the funnel model of innovation yields unsatisfactory results in most cases.

Fig. 5.5 The value of a composite system

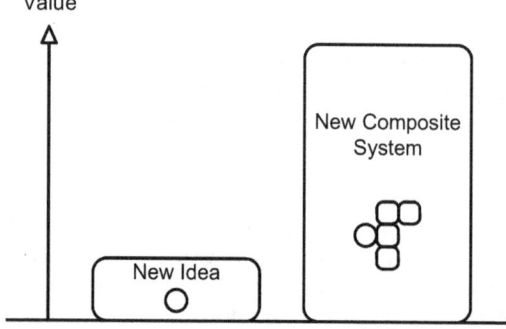

This theoretical result is entirely consistent with the above-mentioned realization that, in practice, it is typically the overall system that generates the commercial value and not the novel element per se. The dependence of the value of an idea from its context can best be illustrated with the use of an example that is drawn from the book *Innovation Factory* that we have already cited (Mock et al. 2013, p. 128):

Example
Let's assume that we concocted a new type of yoghurt in our kitchen at home and that it was an immediate delight for everyone who tasted it. The new yoghurt is, in a sense, our "idea." Now, what is the value of this idea? At first glance, the value seems to be fairly high because it has the property of immediately delighting those who eat it. But how much, really? In monetary terms? This will naturally depend primarily on the quantity of yoghurt that can be sold.

1. *As a private individual: As a private individual, you can convince your family, friends and acquaintances. While you don't wind up earning real money, you do gain a reputation for having certain culinary skills.*
2. *As a dairy owner in a village: As a dairy owner, you can convince all of the village residents and indeed boost your sales. Depending on the size of your village, the added sales may be considerable. You may even have an opportunity to boost your sales even more as a result of the tourists who may begin visiting your village on account of the excellent yoghurt.*
3. *It is a different story altogether if you have concocted the new yoghurt on behalf of a major food corporation. Owing to the global sales opportunities, the potential ranges up into the multimillion bracket!*

What emerges so clearly in the present example is how the value of an invention (idea, product, patent, etc.) crucially depends on the business context. As Chesbrough postulates, it becomes clear that there is no such thing as "the value of an idea in and of itself." On the contrary, it is the business context that determines the value.

> ▶   According to Henry Chesbrough, the value of an innovation is not
> defined by the idea itself, but by the business context in which the
> idea is introduced (Chesbrough 2003, p. 64).

The difference in the value generation for the same yoghurt in three different contexts (home, dairy and corporation) is illustrated in Fig. 5.6.

Let's now recall what we actually set out to achieve by the end of our evaluation-and-selection phase. We wanted to determine which of the many possible innovation ideas was to be selected for further processing in the (expensive!) development phase—and this naturally on the basis of the targeted commercial value. It follows that the scope of our evaluation has also to include the commercial context. This, in turn, means that instead of running through a straightforward ideas competition, we are required to determine the commercial value of ideas embedded in various business contexts. We therefore need to reconfigure the innovation

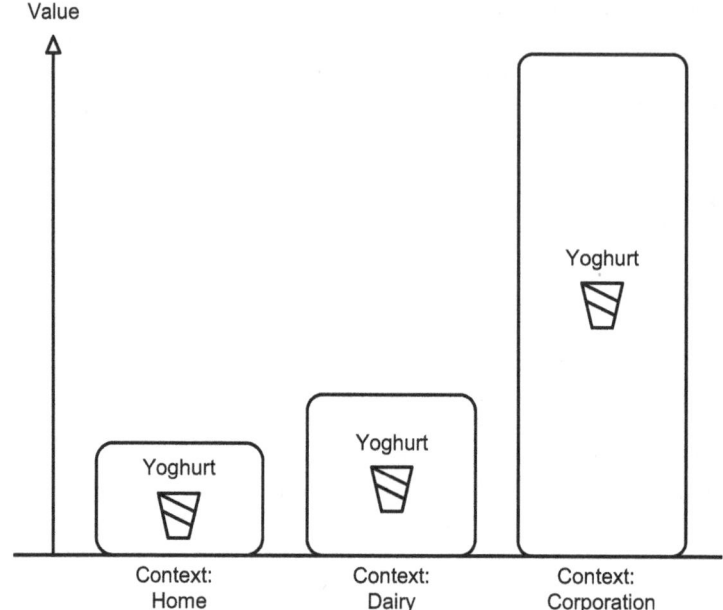

Fig. 5.6 The yoghurt example

process by replacing the previously relatively simple evaluation-and-selection phase with a new and highly complex process!

▶ It is necessary to include the business context in the evaluation of innovative ideas. This means that the commercial value of ideas needs to be determined in relation to various business contexts.

To illustrate this conclusion, we place a new composite system in two different contexts (A and B). As shown in Fig. 5.7, the composite system generates far greater commercial value in Context B than in Context A.

Here, it is altogether possible, and also rather common in the real world, that an idea that is deemed to be little attractive on the basis of a pure cost-benefit analysis turns out to be extraordinarily valuable in a certain business context. It is also often the case that certain ideas have far greater potential to generate commercial value in the context of another enterprise! This can be illustrated using a version of our bar diagram (see Fig. 5.8).

In such cases, the idea should be sold or licensed in this optimal context after it has been appropriately developed. The preparation may include measures such as patenting or the formation of a startup to develop the idea, and that could then be sold at a later point in time.

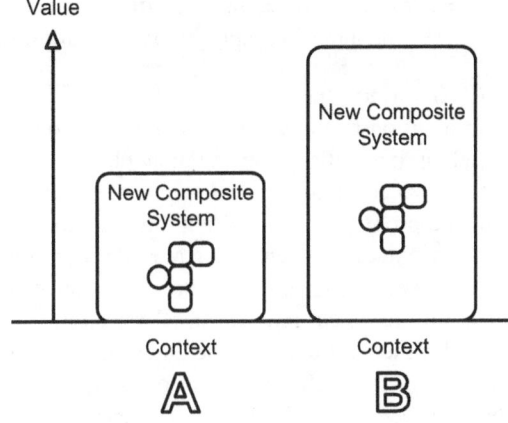

Fig. 5.7 Context-dependent value creation

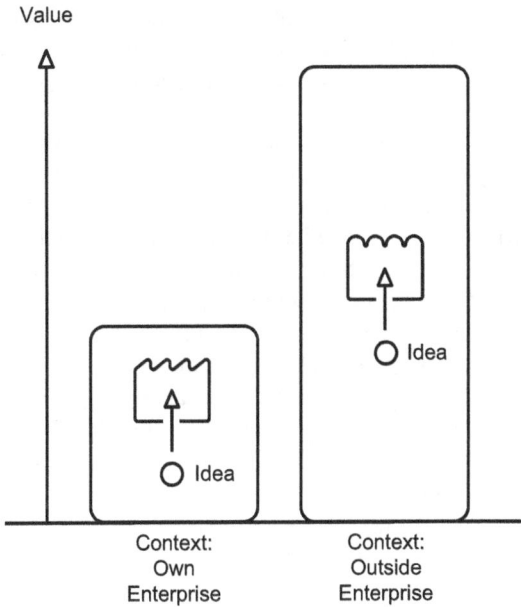

Fig. 5.8 Value creation outside of ones own enterprise

▶ Ideas often realize their greatest commercial value outside of the original enterprise.

This new and complex process involves a veritable exploration of the various possible business contexts. For this reason, we refer to it as a process of "**exploration**." The evaluation-and-selection phase is therefore effectively replaced by a new *exploration phase*. In the following chapter, we turn our attention to a detailed discussion of the contents of this exploration phase.

▶ Exploration: an examination of the various possible business contexts,
   with the aim of finding the optimal business context for an idea.

(b) Ideation phase:

In addition to the evaluation-and-selection phase, the first phase in the classic
model of innovation, the ideation phase, also proves inadequate in real-world
settings. According to the funnel model, appropriate methods are to be used to
generate promising ideas. What this fails to take account of, however, is that it is
often not a matter of generating new ideas ex nihilo, but of finding out what is going
on in the world that could be relevant for the enterprise. Our experience shows that
the ability to detect relevant occurrences in the world has a much greater significance for the well-being of the enterprise than the ability to generate truly original
ideas. And this ability to detect relevant occurrences is simply different than the
ability to be creative in some purer sense.

▶ Experience shows that the ability to detect relevant occurrences in the
   world has a much greater significance for the well-being of the enterprise than the ability to generate truly original ideas.

Enterprises are called upon to acquire the capacity to identify developments that
are potentially relevant for the enterprise. After this, it is a matter of analyzing such
input and interpreting the input from the perspective of the enterprise. Only then
will it be clear whether the new input is at all relevant to the enterprise or can be
safely ignored. Similarly, it won't be clear until after the interpretation is made
which business units within the enterprise are at all affected by the new information
and how they are to respond to the information.

Taking a step back and examining the new capabilities described above lead us
to draw a comparison between enterprises and living organisms. Organisms, too,
need to react appropriately to new information in their environment. Such
organisms also need to be able to distinguish between relevant and irrelevant
information, and be in a position to ignore irrelevant information. Pursuing the
analogy further, the above mentioned new capacity for the enterprise corresponds to
the function of an organism's sensory organs, including the necessary initial
processing of sensory stimuli by the brain. This analogy helps us to grasp why
this capacity is more important for the well-being of the enterprise than the capacity
to generate truly novel ideas. After all, powerful sensory organs are far more
important to the survival and well-being of living organisms than creative output.
And it is not for nothing that powerful sensory organs developed far earlier in
evolution than brains capable of creativity.

▶ Powerful sensory organs are far more important to the survival and well-
   being of living organisms than their capacity for creative output.

Just as in the case of organisms, it is also important for enterprises to detect relevant occurrences or developments as early as possible. This gives one more time for smart responses. A military analogy is helpful in this context. The ability to detect the movement of enemy aircraft as early as possible is extraordinarily important in air campaigns. Air forces rely on early warning systems to accomplish this task. These systems include more than sensory organs (i.e. radar), but also the rapid processing and interpretation of the radar data. It appears similarly appropriate, when considering the newly reformulated phase of innovation and the capacity it calls for, to speak of an early warning system. Here, too, it is not only a matter of detecting something new, but of being able to quickly determine what information is relevant. We therefore also refer to our new and decisive capacity as an **early warning system**. We use an image of a radar screen with the enterprise positioned at its center to illustrate our early warning system (see Fig. 5.9).

Creativity can naturally also be very useful. This also applies in the animal kingdom as well, and evolution has brought forth complex brains to manage creativity. It appears that a capacity for creativity represents an advantage when it comes to survival and evolution. According to our analogy, this should also apply to enterprises. And we know from experience in the real world of enterprises that this is indeed the case. Methods of creativity can and should therefore remain a part of the first phase, although they should be supplemented by an early warning system.

When we now take a look at our expanded first phase of the innovation process, we see the analogue to "sensory organs," "processing and interpretation" and "creativity." From this point of view, we can also extend our early warning system to all three parts of our newly recalibrated process phase. After all, an early warning system outfitted with creativity is simply a more sophisticated early warning

Fig. 5.9 The early warning system

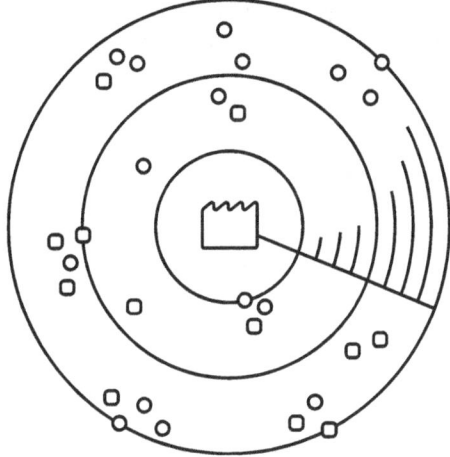

system. We therefore refer to the former ideation phase as the early-warning-system phase.[6]

▶ Early warning system: the analogue to "sensory organs," "processing and interpretation" and "creativity."

(c) Development phase

As described above, the development phase is extremely important for the enterprise. Indeed, the development of a new product represents a major challenge for the enterprise. This applies both with respect to the necessary resources and the inherent risks. For this reason, great attention has typically been paid to this process phase for many years. In the early days of innovation research, innovation was even thought of as synonymous with development. It wasn't until later that it began to make more sense to circumscribe the extension of the term development and to introduce other phases.

The literature available today on the subject of the development phase is extensive and can be regarded as largely complete. Although we need not address the subject any more in the present book, we all the same can expect ever new contributions to still be made to our understanding of the development phase. Despite these aspects, however, we can regard the development phase as sufficiently understood.

▶ Given its comprehensive treatment in the literature, we can regard the development phase as sufficiently understood.

(d) Market-introduction phase

While the same essentially applies to the market-introduction phase, the attention this phase of the innovation process has received has been more recent than that of the development phase.

In the beginning, the market introduction phase was primarily the domain of marketing departments. While the nature of marketing can be regarded in this connection as largely understood, it is to be seen as a necessary, but not a sufficient condition for a successful market introduction. In particular, the works of Geoffrey Moore (e.g. Moore 1995, pp. 3–26) offer a look at the overall context and show that in addition to marketing, important strategic considerations are essential for a successful market introduction.

---

[6]Note: Daniel Huber introduced the term "early warning system" in the context of his enterprise work and discovered that its meaning was intuitive and that its use was well-accepted by both managers and employees alike.

▶   In addition to marketing, important strategic considerations are essen-
    tial for a successful market introduction.

The Redesigned Innovation Process

Figure 5.10 offers a look at the redesigned innovation process.

The process begins with the early warning system. This system represents the sensory organs of the enterprise and is responsible for the detection and initial interpretation of new information. When equipped with a creativity element, the early warning system keeps the enterprise informed of new potential for business development.

We examine the business opportunities associated with this potential in the context of a further step in the process. The exploration phase now delivers specific business opportunities for the first time. Here, the entire business concept is worked out into a rough-and-ready form for each of these business opportunities. In addition to functionality, this includes the commercial concept with the business context and the accompanying business model. The business opportunities will thereby have been roughly, but also comprehensively ascertained. Given that the business context can also be considered for the first time, it is therefore possible for the first time to ascertain the commercial value of these business options. The projects necessary for the development can now be selected on the basis of the commercial value of the business opportunities. Figure 5.11 again illustrates the two central elements of a business opportunity. These include the interplay and the sum of an innovative product concept and a commercial concept consisting of an optimal business context and a future-ready business model.

▶   In   addition   to   functionality,   business   opportunities   include
    the   commercial   concept   with   the   business   context   and   the
    accompanying business model.

The next step is the actual development of the market offer and preparation for the market introduction in a manner corresponding to the described theory. The redesigned model of the innovation process with the yield of the individual phases now appears as represented in Fig. 5.12.

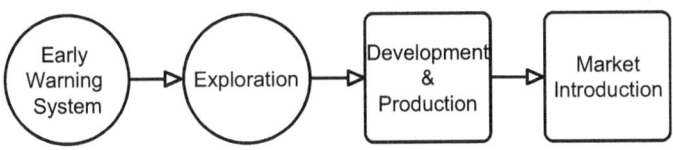

Fig. 5.10 The redesigned innovation process

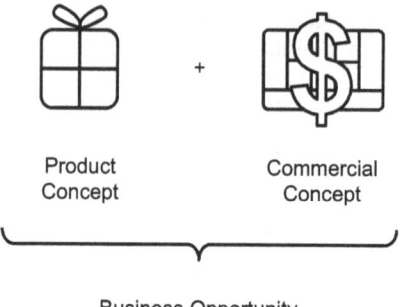

Fig. 5.11 Elements of a business opportunity

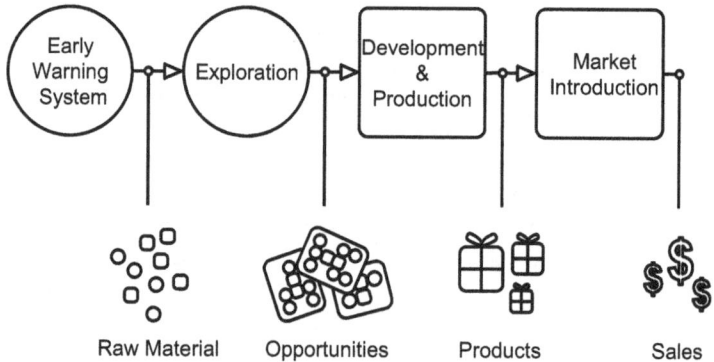

Fig. 5.12 The yield of the individual process phases

Conclusion

We can understand the redesigned innovation process as follows:

As a principle, one needs to first prepare an innovation before it gets possible to implement it. It's only after these two fundamental steps that the innovation then can be brought to the market. The basic innovation process thereby consists of the two main phases of innovation preparation and implementation (cf. Fig. 5.13).

The preparation involves two steps. One has to know where there is potential for innovation and one has to know what exactly the new input means for the business in question: where and how can one integrate the new input in the existing business? As we learned above, this corresponds to our early-warning-system and exploration phases.

The subphases of the implementation phase are also well known. These subphases correspond to the subprocesses development (including the development of any necessary production infrastructure), production and market introduction.

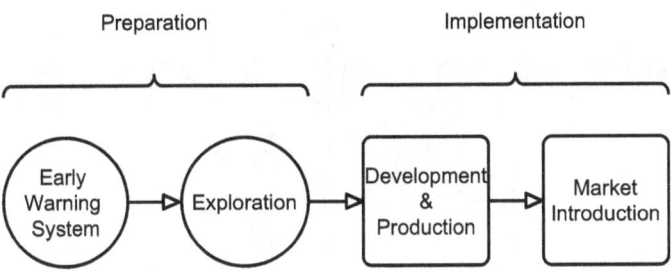

Fig. 5.13 Preparation and implementation

The early warning system delivers the potential areas of business development, or raw material, which are then developed into various business opportunities in the exploration phase. At this juncture, the business management or the supervisory board needs to decide which of the various opportunities are to be forwarded to the expensive implementation phases known as the development, production and market-introduction phases. Given that the opportunities are already available (or must be available) in the form of **business cases**, this decision can be made on the basis of an assessment of the commercial value involved. In the case of very novel proposals, such assessments may however be subject to a considerable degree of uncertainty. Owing to the need to allocate significant enterprise resources, this decision is not free of risk and needs to be carefully considered.

It is now possible for the first time to draft a detailed implementation plan. Ideally, the development and market introduction should proceed according to the plan because any departures from the plan may entail significant cost overruns. Major revisions in the goal can lead to project failure or even to the bankruptcy of the entire enterprise.

The development phase delivers the new products and it is not until their market introduction succeeds that the targeted sales can be generated.

5.3   Transfer

At the beginning of the present chapter, we learned that we need to introduce a key account management to rejoin the enterprise's innovation unit, which was sequestered on account of a difference in enterprise culture, to the rest of the enterprise. We determined that the two enterprise units don't speak the same language and therefore need the help of an interpreter in the form of the key account management. Figure 5.14 offers an illustration of the solution to this problem.

Experience shows that the communication difficulties typically take the following form[7]:

---

[7]An intuitive description of the same effect is also available in Chesbrough (2003, p. 32 f).

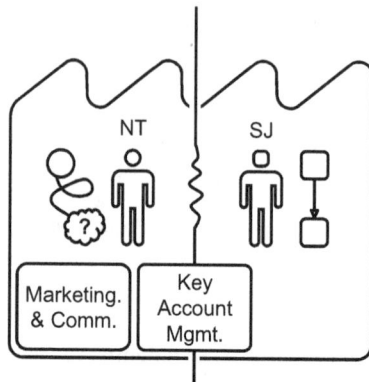

Fig. 5.14 Key account management, marketing and communication

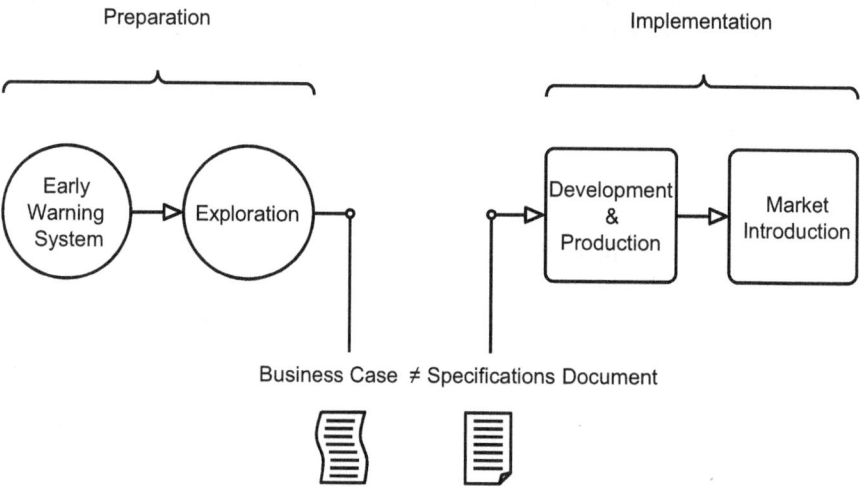

Fig. 5.15 Business case and specifications document

The enterprise's innovation unit submits its final result, the new business opportunity, in the form of a business case to the management. If the management decides in favor of the proposal, it becomes the responsibility of the development department to advance the project. However, the development department is essentially prepared and accustomed to approach its work on the basis of a functional specifications document. The submitted business case can hardly double as a specifications document and is therefore difficult for the development department to understand. Figure 5.15 shows the difference between the business case, submitted at the end of the exploration phase, and the specifications document that is needed as a guide at the beginning of the development phase.

▶ The development department needs a functional specifications docu-
    ment at the beginning of the development phase to guide its work. The
    business case drafted during the exploration is hardly a workable substi-
    tute and is not readily accessible for the development team.

This predicament is typically followed by an exchange of accusations:

- The members of the development team demand a specifications document and
  insist that the exploration unit is far from having completed its work. Moreover,
  they have a certain disdain for the approach the explorers take to their work. In
  their view, the explorers are not really taking work seriously and are only willing
  to get involved *if it's fun.*
- The explorers insist that the submitted business case offers a sufficient
  clarification of what's involved and the ball is now in the court of the developers,
  especially given the fact that the management has made a decision. The
  explorers regard the developers as stubborn and mechanical in their way of
  thinking.

The fact is the developers are not really sure what they're supposed to do with
the submitted business case. The explorers, on the other hand, would indeed be able
to turn it into a functional specifications document. That being said, they're likely
not to know why anyone would need such a document and what it's supposed to
look like. What's more, they tend to truly abhor the kind of work it would take to
draft a specifications document. The result of the conflict is an impasse that leaves
the project on hold. Then the development team proceeds halfheartedly with the
activities (so as not to leave the impression that it has refused to work), leaving the
project to suffocate. The members of the exploration team are frustrated and explain
the failure in terms of the stubbornness on the part of the development team and the
not-invented-here syndrome. And neither the exploration team nor the development
team wants the management to get wind of the failure. They do their best to sweep it
under the rug. This is the death that marks the end of a majority of all innovation
projects submitted from the fuzzy front end. Indeed, experience shows that nearly
all of these innovation projects die this way!

So what can be done to change the situation? For starters, two things need to
happen:

1. The key account management steps in at the outset in the capacity of an
   interpreter so as to make sure that communication between the groups is
   maintained despite their cultural alienation.
2. Someone has to turn the submitted business case into an adequate functional
   specifications document, making sure while doing so that all of the parts of the
   business case are taken into consideration.

As explained above, the tasks and roles of the key account management correspond to those handled in an ordinary sales department. These have been examined

extensively in the literature (see Capon 2001). We will therefore refrain from addressing this subject any further at the present juncture.

The question also arises as to who is to be assigned the job of drafting the functional specifications document and how exactly this work is to be performed. As described above, it is typically the case that the developers "*are not really sure what they're supposed to do with the submitted business case. The explorers, on the other hand, would indeed be able to turn it into a functional specifications document.*" This allows us to conclude that it is the explorers who will have to assume responsibility for drafting the specifications document. However, given that "*they're likely not to know why anyone would need such a document and what it's supposed to look like,*" the exploration team must be given a well-defined order from those who know best, i.e. from the development team.

In the end, it will be necessary for the development team to voluntarily issue an appropriate order to the exploration team for a functional specifications document. Given that the work involved may be extensive, the development unit should be required for paying for that order. This is the only way to ensure that the development team will not insist on a scarcely achievable degree of detail.

▶   The development unit must send the exploration unit a clear order for
       the drafting of a functional specifications document. Given that this may
       entail considerable work for the exploration unit, the development unit
       should be required to pay for this service.

This exchange between the exploration and development units represents an additional phase in the innovation process. We refer to this new phase as the **transfer phase**.

The addition of the transfer phase effectively completes our redesigned model of the innovation process. This model consists of the following three main phases: preparation, translation and implementation,[8] and it includes a total of five process phases, including the early warning system, exploration, transfer, development and production and market introduction (see Fig. 5.16).

The translation work during the transfer phase is so important because the responsibility the NT types have during the exploration phase unit needs to be transferred to SJ types for the development phase. When managing this transfer, it is important to make sure that the knowledge acquired in the exploration phase is transferred as completely as possible in a suitable form to the development unit (see Fig. 5.17).

---

[8]Oliver Gassmann and Philipp Sutter break down the innovation process similarly into two fundamentally different phases. What is referred to in the present book as "preparation" is referred to by Gassmann and Sutter as the "cloud phase." Moreover, they refer to our "implementation phase" as the "building-block phase." Gassmann and Sutter do not, however, describe a distinct transfer phase. They identify the various related activities using different terms and assign them in a different manner to their process elements (Gassmann and Sutter 2008).

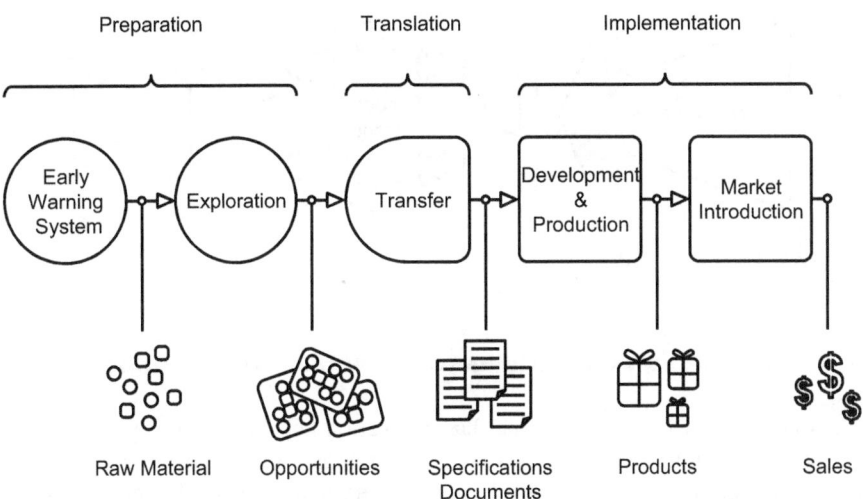

Fig. 5.16 The complete model of the innovation process across five phases

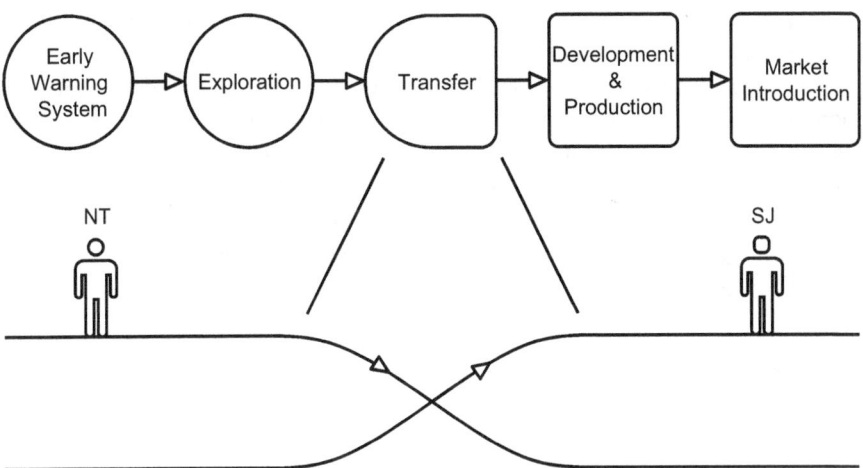

Fig. 5.17 Project transfer in the transfer phase

As mentioned above, business divisions tend to react sensitively to additional costs. Enterprises often agree to fund a percentage of the costs associated with transfer orders so as to encourage development units to pursue transfer orders. Here, it is important not to make it too easy for the business units. They should feel the pinch of the transfer costs, but not be discouraged from submitting orders. The optimal percentage is likely to vary from enterprise to enterprise.

Fig. 5.18 Primary and secondary trends

5.4   Innovation Capacity as the New Main Differentiator

Let's now take a look at the value that innovation has for enterprises. This is an important question when it comes to the organization of innovation. Organization is known to require trade-offs between contradictory requirements. Depending on the value of a requirement, it may be necessary to change existing organizational delimitations to secure an optimal organizational structure. This question concerning the value of a requirement that is to be met by the organization can only be answered on the basis of an analysis of an enterprise's general business environment.[9]

What does the business environment look for enterprises in general? And, even more important, in what ways is this environment changing? To answer this question, we need to take a look at the trends. These, it turns out, come in two different types. We refer to the actual trends as primary trends and to the trends that are triggered by these primary trends as secondary trends. Figure 5.18 illustrates the exposure of enterprises to the primary and secondary trends that arise in their environment.

Primary Trends

Given that a detailed examination of the latest trends would take us too far afield of our subject, we will restrict our discussion of primary trends to a consideration of the following two (uncontroversial) primary trends[10]:

---

[9]The subject of the present book is not a particular enterprise, but enterprises in general. For this reason, we wish to focus only on general market developments and their consequences. We will then be in a position to derive basic organizational principles from the results of our general analyses. When attempting to apply these principles to a specific enterprise, it may naturally be necessary to take account of the enterprise's specific context.

[10]It is clear that one could write entire books about such trends alone. For our purposes, however, it will suffice to consider a few basic developments. The primary trends mentioned in the present context offer an explanation for the accelerating rate and the increasing extent of the changes in the markets. And it's these aspects that are crucial for innovation (They also, by the way, can be used to explain diminishing solidarity in our societies).

- Market globalization
- The establishment of an information and knowledge society

Market globalization is perhaps the change that has had the greatest impact on enterprises in general. In the age of the Internet and very low transport costs, decisions made by competitors located somewhere else in the world can take on an immediate relevance. For instance, it is immediately relevant to a European company when a Chinese competitor introduces a new product to the market. In the age of the Internet, everyone in the world is just a click away.

The rise of the information and knowledge society is reflected in the fact that economic activities have become ever more dependent on information and knowledge. At the same time, it has become ever easier to gain access to information and knowledge. This trend has also been significantly reinforced by the Internet. Private individuals and all of an enterprise's competitors throughout the world have immediate access to relevant information. This makes it imperative for every enterprise to identify and process this information as quickly as possible. An insufficient capacity to do this entails an immediate disadvantage.

Secondary Trends

From these primary trends, we can deduce the following secondary trends:

- It has become increasingly easier throughout the world to gain access to raw materials, goods and services, knowledge, information, labor and capital. For instance, steel factories in Shanghai (where there is neither iron ore nor coal) import iron ore from Brazil and Australia and coal from Inner Mongolia. Apple Corporation in the United States as another example has its products manufactured in China and sells them in the United States and Europe. And global financial markets are the epitome of global.
- The competition for market share has stiffened because geographically distant companies may nonetheless be *present* in ones local or regional pool of competitors. Examples of the irrelevance of geography include the solar industry, the automobile industry and the telecommunications industry.
- The enhanced availability of knowledge and information has led to an intensification of the competition for technological supremacy. This, in turn, has led to accelerated rates of development and to the more frequent introduction of new technologies. Copying technologies has also become far easier.
- The more frequent introduction of new technologies and the intensification of the competition for market share have led to shorter product lifecycles.
- Favorable economic developments have led to increased standards of living for a larger percentage of the world's population (especially in Asia). This, in turn, has led to a situation in which an ever increasing share of overall consumption is accounted for by non-essential commodities (durables). This has resulted in greater consumption volatility, as some consumers are able to simply postpone

the purchase of durables in difficult times. They wait out downswings by postponing the purchase of new vehicles and other products, as was the case, for instance, in the wake of the financial crisis of 2008.

• Greater competition and the facilitated exchange of information and goods have led to an increase in the speed and volume at which money flows.[11]

We can summarize our discussion of trends by concluding that enterprises today find themselves in an environment in which it is becoming ever easier to carry out individual business activities, in which the barriers to conducting business are receding and in which everything is continuously moving faster. In short, the conduct of business has become ever more transitory and complex. The driver for this is a massive increase in the complexity of the business world as a result of an ever increasing degree of interconnection. The pace of this development will of course vary from business sector to business sector.

▶   The conduct of business has become ever more transitory and complex.

Differentiation in Increasingly Ideal Markets

What impact do these developments in the overall business environment have on individual enterprises? In general, we see the following factors:

• Enterprises have an increasingly open access to raw materials and semifinished products. Their dependence on specific suppliers is declining. This situation presents a striking contrast to the beginning of the twentieth century.
• The professional competence of enterprise managements is on the rise, i.e. based on the increased availability of knowledge and on a higher level of academic training. It is common today, for instance, for managers to have MBAs.
• The management of processes, especially production processes, has improved significantly. The drivers here include the increased availability of knowledge and higher levels of employee training.

The result has been an increase in the number of optimized enterprises that perform their tasks ever more proficiently. Given that it has become easier to ensure higher levels of professional competence for the various enterprise functions, the barriers that used to make it hard to enter the competition tend to be lower today. To

---

[11]This has led to a decline in the significance of monetary assets compared to earnings. After all, sufficient earnings will enable one to easily obtain capital. (This represents a striking contrast to the situation in the Middle Ages when the assets of the nobility represented the critical parameter and earnings were relatively less significant.) This constitutes a fundamental change that weakens established enterprises and strengthens new enterprises. The competition thereby becomes stiffer and the rate of change increases.

some extent, you can essentially buy your way in by hiring suitably qualified employees or outsourcing tasks. Resources become more exchangeable, which increases the intensity of competition even more.

What should enterprises do in this increasingly intense and dynamic environment? From a theoretical perspective, the answer is clear. Enterprises need to differentiate. Otherwise, they will go under in a cutthroat competition for ever dwindling profit margins.

Let's then have a look at the possible differentiators. Classic differentiators, for instance, include:

- Access to raw materials: the main differentiator in ancient times and during the colonial period
- Access to capital: the main differentiator at the beginning of industrialization
- Production technology: the main differentiator before the First World War
- Production capacity: the main differentiator after the Second World War
- Marketing: an important differentiator in developed countries since the 1950s[12]
- Market access and distribution networks: an important differentiator since the 1960s (including the less developed countries)
- Sales competence, structures and systems (e.g. local sales outlets, CRM systems,[13] online sales platforms, etc.)
- Quality
- Delivery capacity and customization

The last three items mentioned are more recent differentiators.

Two factors stand out when we analyze the situation today. First, while all of the identified differentiators may continue to play an important role, their relevance will vary depending on the industries and markets involved. Second, the classic differentiators are steadily losing relevance. In other words, it is becoming ever more difficult for enterprises to sustainably differentiate themselves from their competitors.

▶   It is becoming ever more difficult for enterprises to sustainably differentiate themselves from their competitors.

If we extrapolate from these findings with reference to the above-mentioned trends, it seems obvious that the significance of all of the classic differentiators is eroding at an accelerated rate. They are undergoing a transformation from differentiators to basic conditions of hygiene. While they need to be met to survive on the market, meeting them does not give one a competitive advantage over others

---

[12]Today, this also encompasses branding, public image, positioning (high-end/standard/low-end), as well as the implicit values such as community feeling, prestige and street credibility.

[13]CRM = customer relationship management.

on the market. The pace of this development is also accelerating. In short, the task of securing operational excellence has become easier, and has thereby become no more than a matter of basic hygiene.

▶   The task of securing operational excellence has become easier, and has
        thereby become no more than a matter of basic hygiene.

This is a terrifying conclusion! There are apparently ever fewer effective differentiators on the market. This means eroding margins and disappearing profits. This would entail the rise of ultra-optimized enterprises—with a magician-like ability to swiftly conjure up any given products and services in any given quality and without incurring any costs. And this means dwindling profit margins.

The question arises as to what competition would be like in such a world. Who among the magicians is the best magician when all of the magicians do an equal job of mastering their trade? While it may be difficult at first glance to find an answer to this question, the answer is actually quite obvious. Given that our magicians are all equally good at magic, it will not be possible to determine a winner on the basis of their skills at performing magic. The only difference is the result of their magic, the particular things that they conjure. The basis for making the decision as to who wins therefore shifts from *how* they perform to *what* they conjure!

Let's return from our magician analogy to the state of enterprises today. The question now is: what enterprise is the best, i.e. the most successful, when all are equally well-organized and managed? And the answer is that it no longer depends on *how* in the sense of the classic differentiators, but on *what* novel or innovative products the enterprises introduce to the market.

By way of summary, we submit the following prediction about the future. Enterprises will continue to face a massive decline in the significance of classic differentiators in the years to come. These differentiators will be successively transformed, naturally at different rates of speed, depending on the markets involved, into matters of basic hygiene. This process will lead to a shift in favor of innovative capacity as the all-important differentiator. Ultimately, i.e. when all of the classic differentiators have lost their relevance, it is the capacity for innovation that will remain as the only variable by which to gain a competitive edge. And given that innovation—and the capacity of innovation to enthrall consumers—is unlimited, it can never be relegated to the status of basic hygiene. A capacity for innovation is indeed the ultimate differentiator! This turns innovation into the key competence of enterprises.

▶   A capacity for innovation is the ultimate differentiator!

The Consequences for Enterprises

We have thereby found the reason why the subject of innovation has made its way into the consciousness of all of us who concern ourselves with business, including entrepreneurs, managers, economists and lawmakers. It has apparently happened for a good reason. The question now arises as to the direct consequences for enterprises.

Ever since Michael Porter (1985) introduced the notion of the "value chain," we have known that it represents the main process of enterprises. All other enterprise processes play mere subordinate roles to the value chain. We have also known ever since that enterprises are well-advised to organize all of their activities around this one main process. Figure 5.19 shows an enterprise model that is essentially consistent with Porter's model.

The reason why the Porter value chain represents the main process is that the functions in the value chain have a direct impact on customers, and are thereby responsible for the differentiation of the enterprise on the market—at least for as long as the differentiation proceeds according to the classic differentiators.

What happens now when the differentiation no longer matches the classic model and is instead determined solely on the basis of an enterprise's capacity for innovation? In this case, the innovation process is the deciding factor. That being said, innovation doesn't turn up at all in Porter's original model. Reference is made only to *technology management*, a support process that runs parallel to the value chain. However, it makes sense to equate this technology management process, in general terms, with the process of innovation management. If we now apply our criterion for determining the main process to this new situation, we find that the innovation process also represents a main process. In light of its significance, however, the value chain can't simply cease to apply as a main process.

Added to this is the fact that we are still a long way from a state in which a capacity for innovation is the sole differentiation factor, as the classic

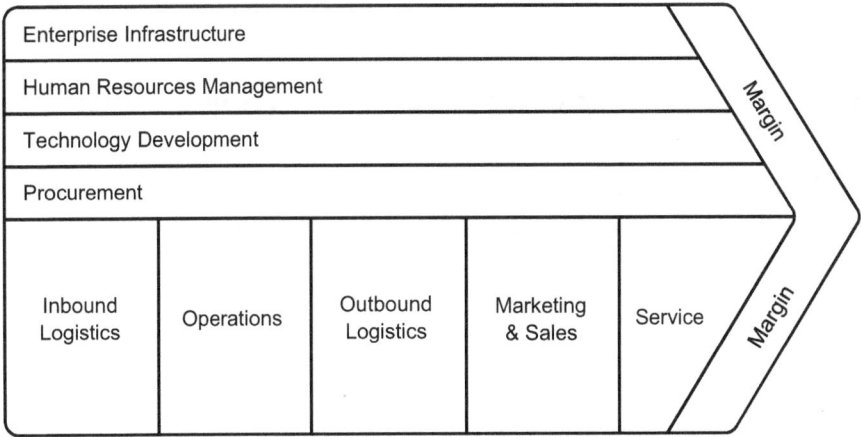

Fig. 5.19 Enterprise model according to Michael Porter

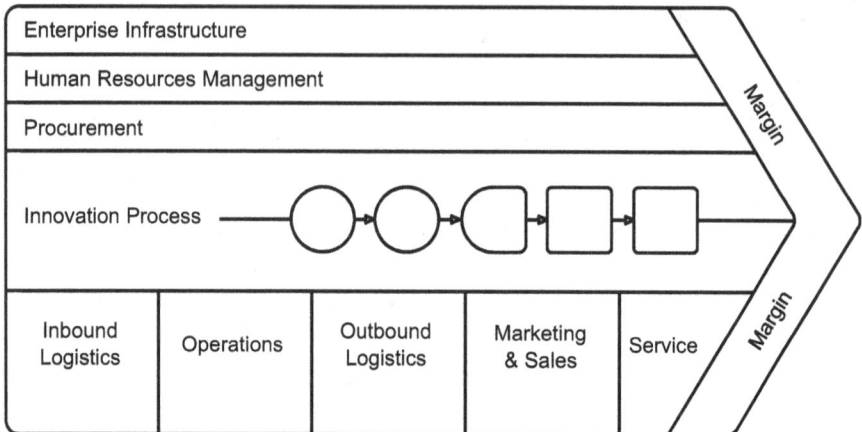

Fig. 5.20 Enterprise model including the innovation process

differentiators continue to contribute to differentiation, although to an ever declining extent. Still, we are moving rapidly forward to a state in which competing enterprises are virtually perfect in terms of their operational performance. This has indeed led to a situation in which the most innovative enterprises have begun to win systematically on the global market. For this reason, we wind up with an enterprise model with two main processes.[14] This adapted enterprise model is shown in Fig. 5.20.

This enterprise model means that we now have to organize the enterprise around two main processes! We address the issue of how to go about doing this in our discussion below.

▶   Enterprises now need to be organized around two main processes: the Porter value chain and the innovation process.

5.5   The Optimized Organizational Structure of the Innovative Enterprise

We have now put together what we can regard as a complete model of the innovation process. You might want to say that we have reached the goal. As we will now see, however, that is not the case. Although our process is accurate in and of itself, it has also saddled us with a number of difficulties. For instance, our

---

[14]This corresponds to the two important levels of "Renewal" and "Optimization" in the St. Gall enterprise model. Please refer in this regard to Rüegg-Stürm (2003) and the *Dual Strategy* by Derek Abell (1993).

experience shows that product development and market introduction are interdependent processes: Depending on the particular product, we may very well need a different approach to market introduction. Similarly, a different approach to the market introduction may call for a different product. If we want to avoid costly learning processes in this situation, we have to make sure that product design and market introduction are developed in a single, holistic process.

▶   **Product design and market introduction must be developed in a single, holistic process.**

However, this means that it will also be necessary to account for both strategic positioning and marketing strategy in the same step. Moreover, the business opportunity can only be appropriately assessed when sufficient general information about all of these factors is available. After all, like different products, different product positions and marketing strategies give rise to different opportunities. And given our innovation process, this holistic design can only be generated in the exploration phase.

▶   **Like different products, both, different product positions as well as different marketing strategies give rise to different opportunities. This makes it necessary to account for both strategic positioning and marketing strategy in the same step.**

Let's take a moment to summarize our progress. We have arrived at a clear, well-structured (and accurate) representation of the innovation process. Upon closer examination, however, we have discovered that it will be necessary to come up with at least a rough-and-ready definition of the entire business opportunity during the exploration phase. This means that it will also be necessary to provide a basic specification for the product design, the strategic positioning and the marketing strategy during the exploration phase, i.e. before working out the budgets.

▶   **It will be necessary to come up with a rough-and-ready definition of the entire business opportunity during the exploration phase. This means that it will also be necessary to provide a basic specification for the product design, the strategic positioning and the marketing strategy during the exploration phase, i.e. before working out the budgets.**

This, however, will have a major impact on hierarchical and organizational structures in the enterprise, changes which cannot be expected to take place without considerable friction. Indeed, power struggles and conflicts are unavoidable. The CEO of the enterprise is the only one who would remain above the fray.

This means that it would only be possible to reorganize the processes involved by taking a top-down approach. The CEO is therefore called upon to personally take charge of the necessary reorganization. It also stands to reason that the preparatory phase of the innovation process (early warning system and, especially, exploration) need to be placed directly under the authority of the CEO. Any other arrangement would be destined to fail on account of the remaining power structures within the enterprise. At the same time, the strategic positioning will make it necessary to secure solid cooperation with those responsible for strategy. It is therefore necessary to upgrade the status of the person in charge of the early warning system and exploration to the executive-management level, reporting directly to the CEO, parallel to the strategy unit. This leads to our first rule on the organization of innovation in an enterprise:

▶   Rule 1: The early-warning system and the exploration unit must be
    located centrally in the organizational structure of the enterprise. This
    central innovation unit is charged to holistically process and evaluate
    innovation projects, and is to report directly to the CEO.

There is, however, one exception to our first rule. Business opportunities that fall exclusively within the domain of an established business division can (and should) be assigned to the respective business division. This is because the business division can be expected to be in possession of the relevant experience and knowledge that will be required to best exploit the opportunity. However, the type of innovation involved in such cases is limited to sustaining innovations (Christensen 1997), a topic that we will discuss in greater detail below. Moreover, the organization of the relevant business division in this exception case is to mirror the structure outlined above. In particular, it is to be outfitted with a specific exploration unit that is directly reporting to the head of the business division and arranged parallel to any staff responsible for strategy. Other organizational forms can be expected to fail in this context for the same reasons that they do in the enterprise at large. It is a matter here of fractal organization. Our second rule for an optimized organizational structure to serve the purpose of enhancing an enterprise's capacity for innovation is therefore:

▶   Rule 2: The business divisions are responsible for innovation in their own
    business domains.

The model organization of an innovative enterprise can be represented via the organizational structure shown in Fig. 5.21 (the exploration and innovation units are referred to in this case as **central innovation**).

We can naturally not exclude the possibility that innovation projects that arise in the context of a particular business might also make sense in some other business

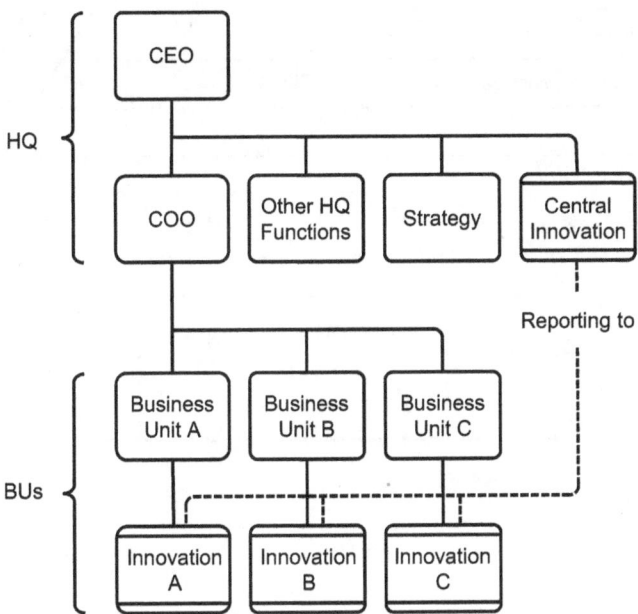

Fig. 5.21 Optimized organizational structure for innovative enterprises

context. This applies especially when one considers joint efforts with other business units or when it might be appropriate to analyze business contexts beyond ones own enterprise. For the reasons mentioned above, such analyses can only be carried out exclusively at the level of the enterprise by the central innovation unit. It is therefore essential to make sure that this unit is comprehensively apprised of all innovation efforts underway in the various business divisions. This leads us to our third organizational rule:

▶   Rule 3: Any innovation units organized within the business divisions are
        to report to the central innovation unit at the level of the enterprise in
        the context of a dotted line reporting relationship.

Final Remarks on the Organization of the Innovative Enterprise

We have now seen how to organize the **Innovative Enterprise**.

In addition to the value chain, the corresponding enterprise model encompasses the innovation process as a second main process. Figure 5.22 offers an illustration of the enterprise model.

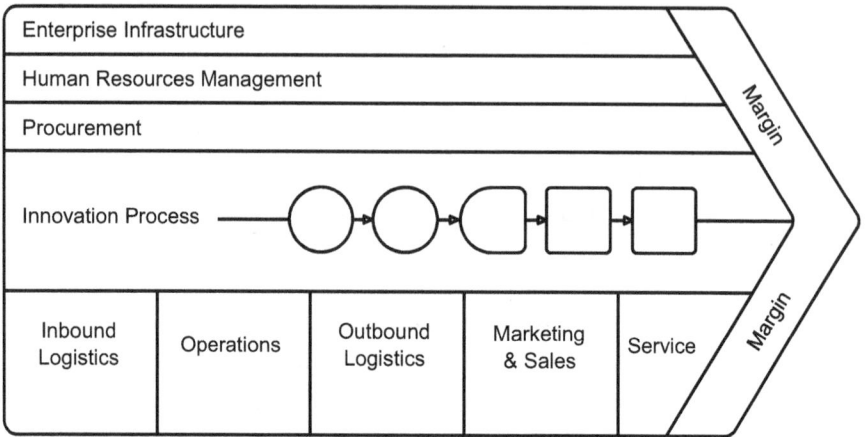

Fig. 5.22 Model of an innovative enterprise

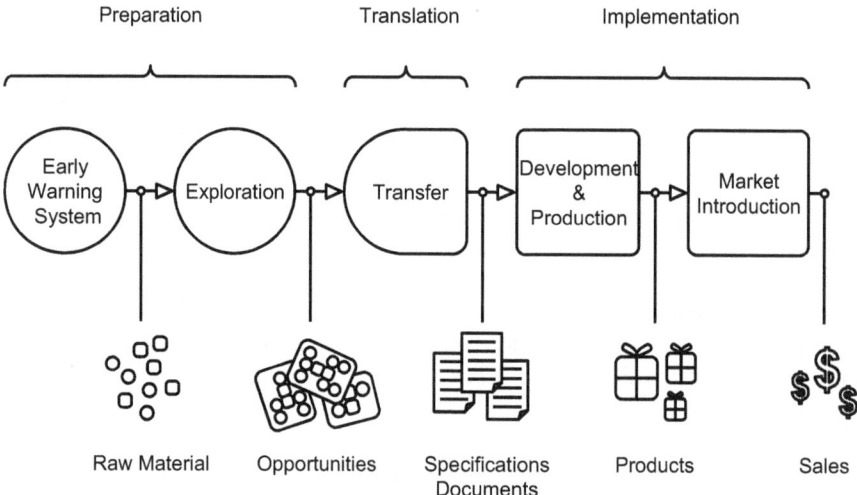

Fig. 5.23 Model of the innovation process

Figure 5.23 shows a model of the innovation process.
Figure 5.24 shows a model of the optimized organizational structure.
Three rules of organization apply:

- **Rule 1**: The early-warning system and the exploration unit must be located centrally in the organizational structure of the enterprise. This central innovation unit is charged to holistically process and evaluate innovation projects, and is to report directly to the CEO.
- **Rule 2**: The business divisions are responsible for innovation in their own business domains.

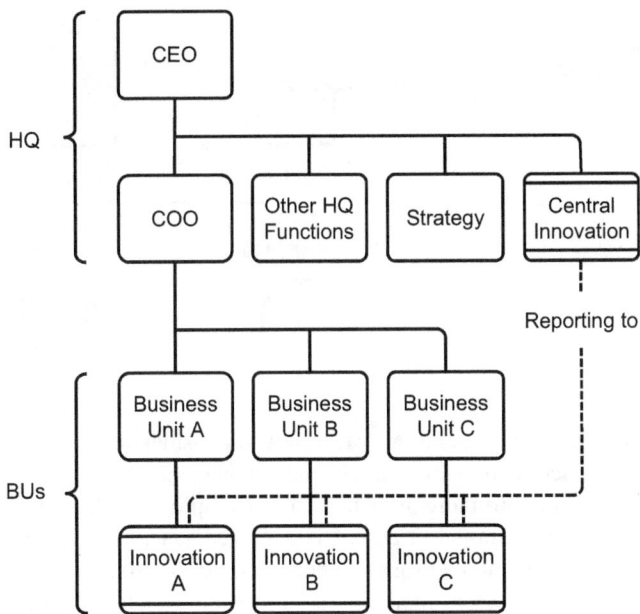

Fig. 5.24 The organizational structure of the innovative enterprise

- **Rule 3**: Any innovation units organized within the business divisions are to report to the central innovation unit at the level of the enterprise in the context of a dotted line relationship.

5.6   The Organization of Innovation in Small Enterprises

We have so far presented our organizational considerations from the perspective of a corporate environment. There are two reasons for this. First, our findings are especially relevant to large enterprises. Second, the organizational consequences can be vividly illustrated in the case of larger enterprises. This is because major changes in large enterprises always go hand in hand with changes in their organizational structures and in their (often sharply defined) processes. Such changes are therefore easy to represent.

The question now arises as to how our results can also be applied to small or medium-sized enterprises. Smaller enterprises have comparatively streamlined organizational structures and the smaller they are, the more general and the more informal their processes tend to be. While one-person businesses at the lower end of the size scale have exactly one organizational unit, namely, the head of the company, and no formal processes, they still need to execute all of the functions of an enterprise, including production, sales and bookkeeping. And given that this

Fig. 5.25 Organizational structure based on time management and employee roles

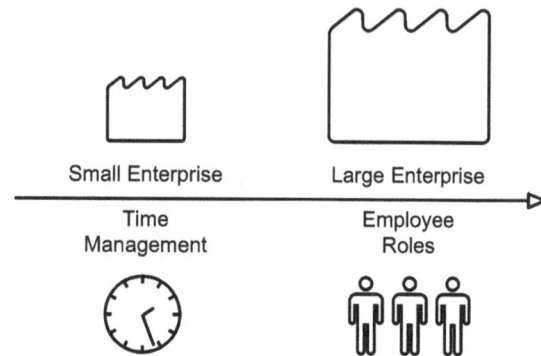

naturally also applies to innovation, we're left to consider what structure is to be adopted. This structure is achieved via time management (in contrast to large enterprises where the structure is achieved via employee roles). In one-person enterprises, the organizational structure is achieved via time management only. Specifically, it is important to ensure that enough time is available to address such a non-operational subject as is innovation. The result can be generalized for all smaller and larger enterprises alike. As illustrated in Fig. 5.25, the smaller an enterprise is, the more likely it will be that structure is achieved via time management. And the larger an enterprise is, the more likely it will be that structure is achieved via employee roles.

▶ Even one-person enterprises need to execute all of the functions of an enterprise.

▶ The smaller an enterprise is, the more likely it will be that structure is achieved via time management. And the larger an enterprise is, the more likely it will be that structure is achieved via employee roles.

When it comes to medium-sized enterprises and exploration units, which are comparatively small even in large enterprises, this means that the function of the central innovation unit will have to be supplemented via time management approaches. This may involve the assignment of one to five employees to a central innovation unit while at the same time limiting many of the related exploration activities to workshops and projects. The small central innovation unit is then responsible for managing these temporally circumscribed efforts.

Smaller enterprises will then be entirely limited to the scheduling of workshops and projects. The temporally limited elements of exploration in this case are coordinated and managed by executive staff, usually by the CEO or an innovation officer in the executive management.

It warrants pointing out at this juncture that time management will naturally also play a role in larger enterprises, although to a lesser degree.

While the conclusions we reach in the present book for large enterprises therefore apply equally to small and medium-sized enterprises, the functional performance will be achieved primarily via time management, as opposed to the establishment of organizational structures in the case of large enterprises.[15]

5.7   The Case of Plenaxx

The case of Plenaxx is an example of how certain constellations can prevent successful innovation despite the best of conditions.

The story begins at the end of 1999 when a startup was formed at the joint initiative consisting of Mobiliar Versicherung, UBS, Swisscom and Valora Logistics. The vision behind this startup was to provide web services to medium-sized companies that would essentially allow these companies to boost their work flexibility. At the time, web portals were the prevailing vision in internet-based business. Plenaxx was to become this sort of portal. The service scope of Plenaxx consisted of a virtual team space (closed user group) that enabled company-wide joint processing and archiving of data with technology that was quite advanced for the time.

Solutions were already in place both for versioning and user notification. Plenaxx anticipated many of the functionalities that would later be so successfully commercialized by Facebook, SharePoint, Dropbox and Doodle.

While the enterprise managed to prepare its services for market introduction, the dot-com bubble burst in the middle of the rollout with a functioning application, the first few thousand customers and steep growth. In the ensuing panic, the owners were not in a position of drawing a distinction between an inflated evaluation of ideas and solid business development. The project was stopped in 2001. At the time, Martin Steinmann was a member of the Plenaxx management, which was comprised of 60 employees before the liquidation.

To put it simply, Plenaxx experienced three life phases that were very different in terms of organizational form, prevailing culture and approach to innovation. In addition to the substantive challenge with regard to the innovative services to be provided, this continual pressure to transform was an additional stress factor.

Phase 1: Euphoria and hopeful anticipation: successful recruitment of future shareholders, successive expansion of the core team, unconventional approach

---

[15]Innovation in medium-sized enterprises is currently often not secured via a certain organizational unit, but by the natural and more intensive interaction medium-sized enterprises maintain with their customers and suppliers in the daily conduct of business.

to management tasks. For instance, empty office space was furnished over night without approval and populated with temporary employees, a practice that was only approved after the fact...

Phase 2: Euphoric growth: enhanced staff diversity by design, including more diversity in terms of their curricula vitae and life histories, also extending to older workers, i.e. not just wild youth. Owing to the exhausted job market, an increasing number of those hired came with a corporate mindset instead of an entrepreneurial mindset. The project turned into an enterprise. A functioning infrastructure for this was naturally a necessity. The number of operational tasks increased.

Phase 3: Euphoric consolidation: the orientation towards market conditions and competitors intensified. The distribution of work became more defined, which led to more pronounced formation of hierarchies and roles. The pledge of allegiance to startup principles became more contrived.

How are we to understand the difficulties? Plenaxx had actually met the prerequisites for success: solid partners with financial clout and established distribution channels, a compelling concept and a compelling product pointing the way to the future. The problem was that the shareholders never really understood the strategic significance and true implications of the business idea. This was especially apparent when the decision was made to abandon the project and the shareholders began to act more like financial investors than strategic partners. The dramatic shift from optimism to pessimism on the larger dot-com market was presumably the trigger for their sudden interest in damage control.

While the product was very promising, it was well ahead of its time and therefore in need of justification. Plenaxx didn't have the courage to approach the market in small steps that would have been comprehensible for the market. It wanted too much all at once. Influenced by the "new economy," the perception of the management was also one-sided. Developers worked intensely on the next release and over interpreted the signals sent by the competition. In the euphoric atmosphere, the enterprise nearly rushed itself to death. Plenaxx was under the illusion that if the competition copied its functionalities that the competition would also understand them all. However, this was generally not the case. While Plenaxx was already running functionalities smoothly, the competition had not yet figured out the copies they had obtained in any detail. In retrospect, it would have taken them months or even years to catch up to Plenaxx.

Plenaxx also qualifies within the terms of our discussion as an Innovative Enterprise in which the exploration, prototype development, demonstrators and transfer worked as is typical of startups. Although much was done right, a wide array of uncertainties and smaller mistakes ultimately sealed the fate of the young enterprise.

Conclusion: This is an example of how projects can fail despite the best of conditions. It is clear that innovative enterprises, and especially startups, go through various developmental phases in short periods of time and are required to adapt accordingly, with each and every change. This exposes them to danger from both outside sources and inside sources. If certain milestones are reached in demonstrations of brilliance, this does not yet say anything about the chances of taking the next challenge in stride. In the case of Plenaxx, various reasons converged somewhat fortuitously to lead the company to a regrettable end.

References

Abell, D. (1993). *Managing with dual strategies: Mastering the present, preempting the future.* New York: Free Press.

Capon, N. (2001). *Key account management and planning.* New York: Free Press.

Chesbrough, H. (2003). *Open innovation.* Boston: Harvard Business School Press.

Christensen, C. M. (1997). *The innovator's dilemma.* New York: Harper Business.

Eversheim, W. (Hrsg.). (2003). *Innovationsmanagement für technische Produkte.* Berlin: Springer.

Gassmann, O., & Sutter, P. (2008). *Praxiswissen Innovlationsmanagement.* Munich: Hanser.

Johnson, M. (2010). *Seizing the white space.* Boston: Harvard Business Press.

Mock, E., Garel, G., Huber, D., & Kaufmann, H. (2013). *Innovation factory.* Fribourg: Growth Publisher.

Moore, G. A. (1995). *Inside the Tornado.* New York: Harper Business.

Porter, M. E. (1985). *Competitive advantage.* New York: Free Press.

Rüegg-Stürm, J. (2003). *Das neue St.Galler Management Modell.* Bern: Haupt.

Exploration

6

6.1 What Is Exploration?

Let's recall what we said in the previous chapter about the new exploration phase that was introduced to replace the evaluation-and-selection phase:

> We wanted to determine which of the many possible innovation ideas was to be selected for further processing in the (expensive!) development phase—and this naturally on the basis of the targeted commercial value. This new and complex process involves a veritable exploration of the various possible business contexts.

This is all based on the notion that the value of business ideas is derived primarily from their business contexts.

The exploration phase thereby consists of the following main tasks:

- Draft appropriate functional solution concepts.
- Demonstrate technical feasibility.
- Draft business concepts, including a general marketing concept.
- Demonstrate commercial feasibility.
- Use the results gathered so far to draft a business case (rough business plan) for each business option judged to be attractive and determine their commercial value.
- Select the most suitable opportunity.

Figure 6.1 shows the main tasks that are to be performed in the exploration phase, as well as their dependence on one another. The figure shows that three

© Springer International Publishing AG 2017

D. Huber et al., *Bridging the Innovation Gap*, Management for Professionals, DOI 10.1007/978-3-319-55498-3_6

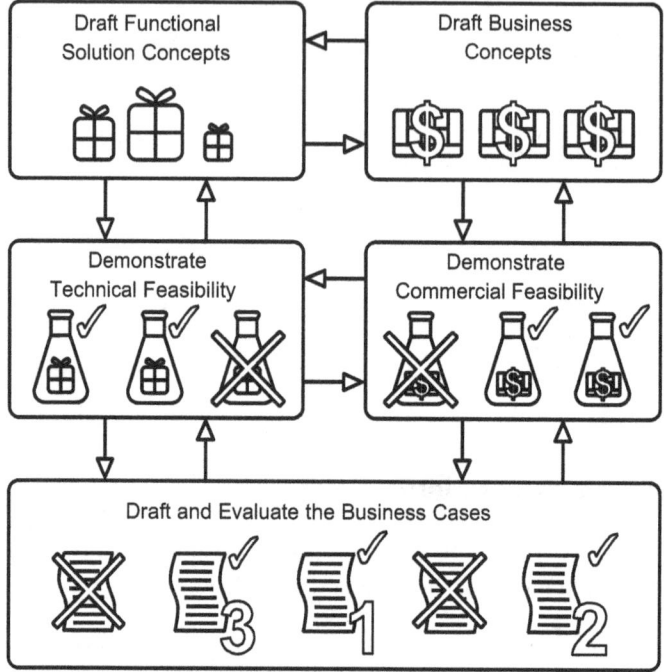

Fig. 6.1 Main tasks in the exploration phase

attractive business cases have been drafted. The third business case was selected as the best opportunity.

The drafting of the functional and business concepts in this context is an especially creative task. The performance of the task thereby involves a high degree of learning and therefore requires an iterative approach.

▶ The exploration phase includes the demonstration of the technical and commercial feasibility. This is a highly creative task that requires a great deal of learning.

Technical iterations (what can we do with this thing?) typically alternate in this context with iterations in which the focus shifts to the needs of the customer (what problems can we solve for potential customers?).[1] However, given that we know nothing at the outset about the product and the customer, we are left only with the iteration between these two poles. Figure 6.2 offers an illustration of the iteration between the two different iteration loops.

---

[1]Mark Johnson refers in this regard to "a job to be done" (Johnson 2010, p. 25 ff.).

Fig. 6.2 Iterations

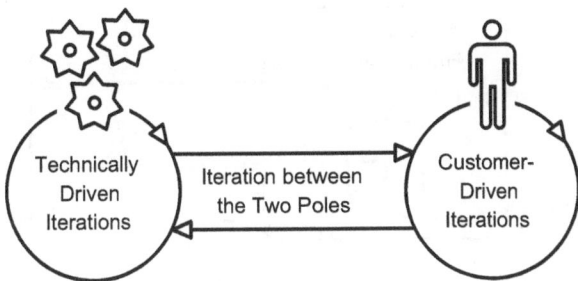

The fact that it is necessary to simultaneously examine many different variations in iteration loops means that the number of examination steps and dependencies between the steps will be very large. It is therefore essential to keep these examination steps as simple as possible.[2] This is why desktop research combined with intensive reflection is the dominant work method. Success in this particular case will therefore not depend on the number of employees assigned to the task, but on their suitability for the task. This also shows why it is so important not to incorporate these exploratory learning phases into the costly development phase. The costs would otherwise skyrocket!

> **Business lab**: As described above, success in the exploration phase depends critically on the drafting of the technical concept in various versions and on the development of a suitable commercial concept: the **business case**. Ultimately, it is this commercial concept that determines the business model and the business context. And it is this business context that is crucial to the commercial value of our innovation. We refer to this process as **business exploration**. In this process, the drafting of the commercial concept always is done hand in hand with the drafting of the technical concept: i.e. the technical exploration.
>
> Processes of arriving at a technical concept are well known. One essentially engages the services of engineers or scientists to solve any problems that arise by providing them with a laboratory in which to test the suitability of the proposed technical solutions. Such technical laboratories are standardly available in enterprises.

(continued)

---

[2]In addition to iteration, two other factors are essential here: (1) There is no such thing in innovation projects as the continuous development of results. Project progress typically takes place in fits and starts. (2) Not everything needs to be actively processed. Sometimes it is important to step back and let the project take its course. As in the case of fermentation, results sometimes develop on their own.

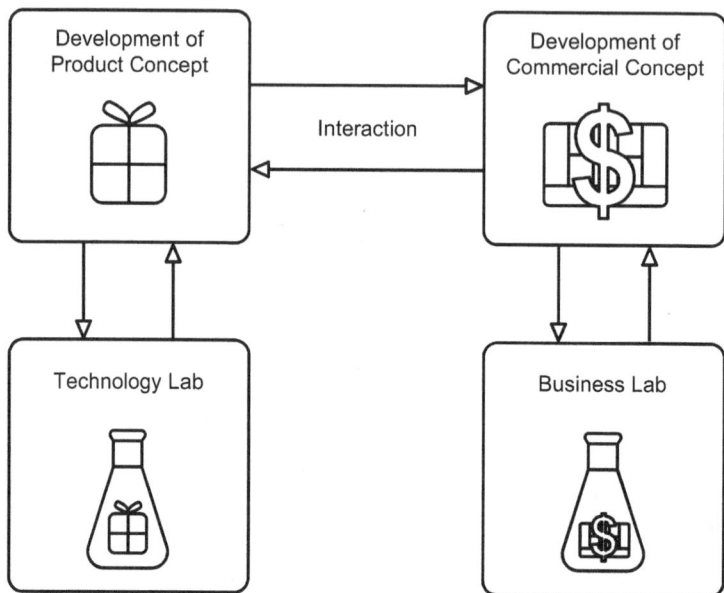

Fig. 6.3 Business lab

The task of arriving at a commercial concept presents more of a challenge. As in the case of the technical concept, we also need trained employees to develop the commercial concept, employees who are in a position to identify suitable business contexts and develop attractive business models. Here, too, we encounter a need to conduct tests and experiments in an orderly framework. Moreover, we also need a secure work environment, a kind of laboratory, which, however, is not arranged according to technical parameters, but to commercial parameters. We refer to this laboratory for the commercial areas of our exploration as a **business lab**.

Figure 6.3 offers a representation of the interactions between the development of a product concept and the technical laboratory (technology lab), as well as the development of the commercial concept and the business lab.

Such business labs are extremely rare in enterprises today. For this reason alone, it is little surprising that so many innovation projects are commercial failures. It is simply a matter of an inadequate upstream examination of the associated commercial aspects. Owing to their intense mutual dependencies, it is imperative for business labs and technology labs to work hand in hand. It therefore makes a lot of sense to ensure their physical proximity.

6.2   Refinement and Decision Preparation: The Explorer's Work Environment

In order to secure an optimal exploration process, we will need to consider one other aspect of successful innovations. This aspect is that they typically involve relatively few elements that are completely new.[3] Most of innovative products are mainly comprised of existing parts and subsystems. The innovative new systems are created by combining new elements with existing elements in a nonprecedented way. The commercial value of such systems may come to exceed that of the original newly created elements by several orders of magnitude. We might even say that the original element has been refined via the process of combination and integration to form a product that is far more valuable.

▶   Successful innovations typically involve relatively few elements that are completely new. Most innovative products are mainly comprised of existing parts and subsystems. The original product is typically refined via a process of combination and integration to form a significantly more valuable product.

This means that one of the main approaches taken in the context of exploration is to combine and integrate various existing as well as new elements to form new composite products. Successful exploration will therefore depend on the availability of a large number of suitable elements and subsystems. Given that we can't know in advance what elements and subsystems we need, we will have to make these available as interactively as possible.

The task of explorers resembles that of children playing with LEGO®. If we expect the explorers to be as creative as possible, then we better make sure they are given many and various kinds of pieces, i.e. product elements and semifinished products. Moreover, children playing LEGO® typically start by spreading all their pieces out in front of them. This allows them to quickly survey all of their pieces and find inspiration in what they see. The creative construction then takes place with the joy of an artisan in a process of looking ahead and testing various combinations of the available elements, i.e. by repetitive experimentation.

In the case of our LEGO® example, there is naturally a degree of disorder and creative license. Real exploration is no different. And the processes that explorers complete on their way to innovations and business opportunities are just as elusive as those of children playing with LEGO®. The only thing we can do is to provide a stimulating work environment and as many different elements as possible.

With reference to athletic contests, let's refer to such work environments for explorers as **exploration arenas**. Like in football, a game does not progress

---

[3]Nassim Taleb has remarked: "The test of originality for an idea is not the absence of one single predecessor but the presence of multiple but incompatible ones" (Taleb 2013, p. 12).

according to a predefined process, but develops largely spontaneously. The ball goes back and forth across the field until things crystalize in the form of a series of plays that lead to a goal. In the next chapter, we introduce a graphic of the arena to better illustrate the processes that takes place in it.

How then should we design an exploration arena?

Project rooms, virtual or real, constitute the arena space in which multiple explorers perform their work and have contact with one another. This setup requires the fulfillment of a few prerequisites:

- Explorers should be able to communicate well with one another in an everyday sense and, in particular, with respect to their common endeavors.
- Explorers should know one another as well as possible. In the case of virtual exploration arenas, it is advisable to give explorers an opportunity to get to know one another in the framework of a kick-off event (socializing).
- The most valuable exchange between explorers tends to be informal. One should therefore establish informal settings to facilitate the exchange of ideas (e.g. a café). This cannot be replaced by formal meetings.
- Audiovisual communication is helpful in the case of virtual teams. It is also helpful to make sure that the connection is left uninterrupted (e.g. a Skype connection that is always on). This permits a certain degree of informality despite the geographic distance.
- Employees working at the same geographical site, should be placed in immediate vicinity to allow that they can take breaks together.

The critical factor therefore is the facilitation of informal exchange.

Excellent information systems play an important role when it comes to the elements that are to be combined or integrated with one another. Provision should be made for a database or a **warehouse for the various building blocks**. This warehouse provides a space for storing concept building blocks and partial solutions. As in the case of LEGO®, it is important not to throw away any building blocks. This is so because one can never know a priori that the concept building blocks that appear irrelevant at the moment will not be the priceless key to a future problem. In contrast to pieces of LEGO®, which need to be disassembled for later structures, exploration building blocks can be left whole and redeployed in multiple copies whenever appropriate. This circumstance permits additional, conflict-free deployment opportunities.

6.3   The Exploration Process

Although we suggested above that "*the processes that explorers complete on their way to innovations and business opportunities are just as elusive as those of children playing with LEGO®*" one should at least attempt to model the exploration process in a generic sense.

The exploration phase can be broken down into two subprocesses. The first subprocess of **refinement** takes place in the exploration arena and the second

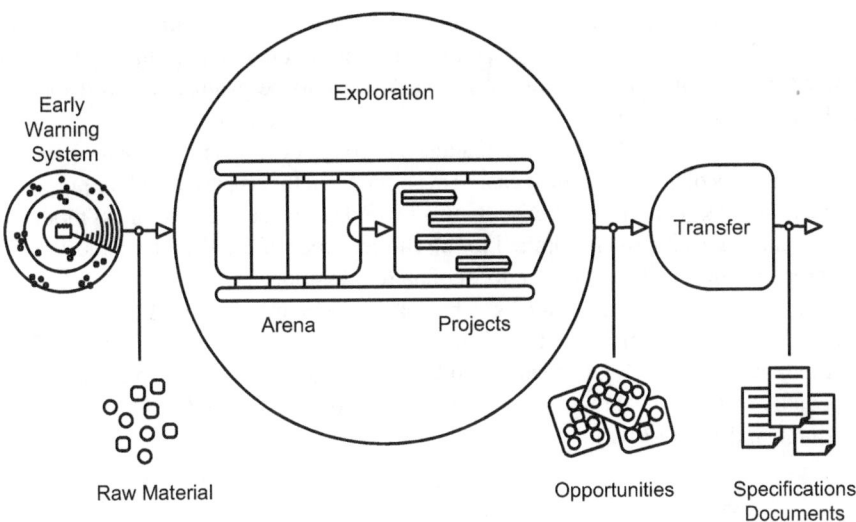

Fig. 6.4 The exploration process

subprocess of **decision support** takes place in the framework of exploration projects. The exploration projects in the second subprocess are compiled in a project portfolio. Figure 6.4 offers an illustration of the two subprocesses of the phase.

Subprocess 1: refinement in the exploration arena: various **concept building blocks** are first brought into the exploration arena. Here, we speak explicitly of concept building blocks instead of ideas. Ideas are often regarded as completely finished elements and are administered in idea management systems in many enterprises. A business opportunity that has been completely worked out and refined, however, includes far more than an original idea. Such fully outlined opportunities are based on many different technical and commercial elements. The term concept building block, or simply building block, is therefore well-suited as a term of reference for these various elements.

The first step is to go to the exploration arena and create an integrated concept out of technical and commercial building blocks. The building blocks available in the exploration arena may be newly created or they may come from the early warning system. As described above, the building blocks are then combined iteratively, augmented or broken down and reassembled with new or existing components. If something is missing, then a search is conducted for the missing building block. If necessary, the missing building block can be created in the framework of research efforts. The processes thereby include: combining, partitioning, augmenting and integrating. This work involves a high degree of complexity and requires a lot of creativity. Once the integrated concept has been established, then it is time to apply the same principles to find the optimal business context within a work arena. This mechanism gives our new concept its value and effectively turns it into a business opportunity. The first subprocess (refinement) takes place in the exploration arena and is a continuous and iterative process and not a project in the strict sense of the term!

Subprocess 2: decision support through an exploration project: business opportunities that have been developed in the exploration arena are then developed further in a second subprocess until they are ready to be evaluated in a decision-making process. Concept consolidation and validation are carried out. The business opportunity is described in the usual manner as a rough business plan. We refer to this rough business plan as a **business case**. This is the form that is best suited for purposes of decision making in the enterprise. In this second subprocess, the business model is refined and a specific plan is drafted for all further procedures. Therefore, the second subprocess will ideally exhibit the form of an (exploration) project.

If the innovation in question is a business-model innovation, the exploration project can only be carried out after the business context has been identified. The methods developed by Osterwalder and Pigneur (2010) and Johnson (2010) are especially suitable to do that. Afterwards, one can proceed according to Ries (2011) or Blank and Dorf (2012).

The Refinement in the Exploration Arena

To illustrate the refinement process in the exploration arena, we use a graph that resembles a football field (see Fig. 6.5).

The representation of the football field underscores the creative and playful aspect of exploration work. Moreover, successful exploration work also requires a lively exchange of ideas among the explorers. As in the case of many team sports, communication and team spirit play an important role.

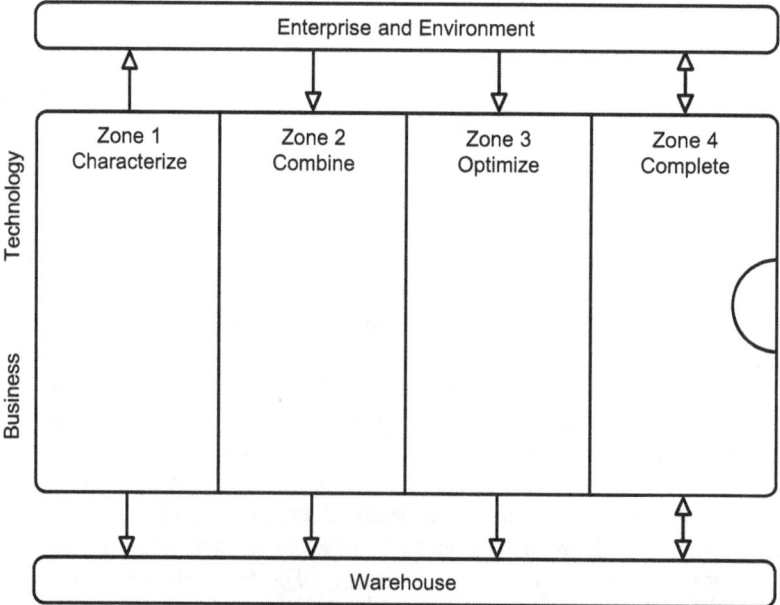

Fig. 6.5 The exploration arena

The playing field exhibits four different zones, including Zone 1: *Characterize*, Zone 2: *Combine*, Zone 3: *Optimize* and Zone 4: *Complete*. The upper area of the arena includes the technological aspects of exploration (technologies, products, systems, etc.) and the lower area includes the business aspects (business contexts, markets, business models, customers, etc.). Depending on the current position of a given concept on the arena's vertical axis, the focus of the exploration will be on technological or business aspects.

The fourth zone "complete" also includes a separate end zone. This end zone is where the developed and finalized business opportunities are positioned, a point from which they can be transferred to the downstream decision-support subprocess. The proposed generic refinement process thereby crosses a total of four zones in the arena and is focused on the development of the most valuable business opportunities from the many technological and business building blocks in a relevant business context.

Zone 1: Characterize

In the first phase, the technological and business building blocks identified by the early warning system are all initially placed in a warehouse (a database or a physical storage facility). Given that the value of a building block is not inherent, but dependent on a context, it will not yet be possible to differentiate between superior and inferior building blocks. Furthermore, building blocks that appear at first glance to be useless may turn out to be very valuable later. For this reason, we recommend that all of the concept building blocks be stored as a matter of course (see Fig. 6.6).

To ensure our ability to locate suitable building blocks as needed, we label them with references to their typical properties at the moment of storing them. Depending on the context involved, these properties may indicate a strength or a weakness. While we are aware of the fact that it is not yet really possible to ascertain strengths and weaknesses because the context is missing, the approach to assign strengths and weaknesses out of an arbitrary context nevertheless establishes these properties in the first place. This nonetheless assures us to a sufficient degree that the essential properties of a certain building block will be identified.

We therefore select an arbitrary context in order to be able to identify the properties in a specific case as advantageous or disadvantageous. This allows us to avoid getting stuck in a realm of non-binding abstractions.

In this first zone, we begin by shortlisting the number of building blocks we wish to actively process by a factor of around two. We simply continue to process the more attractive of the building blocks and set the less attractive aside. This selection is naturally somewhat arbitrary because we haven't yet identified a specific context. For reasons of efficiency, however, we make a selection, and any building blocks that have been wrongly set aside can be inserted back into the process later. The process thereby has a certain self-healing feature.

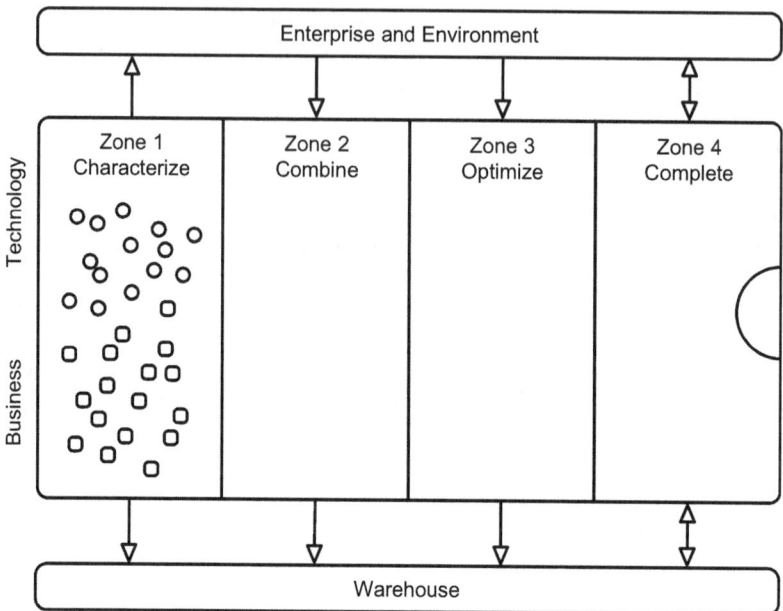

Fig. 6.6 Zone 1: characterize

Building blocks are also identified in this step that can be directly implemented for business purposes without further exploration. Such "Quick Wins" are to be forwarded directly to the relevant business divisions.

Zone 2: Combine

In the second zone, we break the building blocks down into their elements and attempt to combine or multiply the strengths of various building blocks and, on the other hand, to eliminate or compensate for their weaknesses. This process leads to combo building blocks or **building block conglomerates** whose individual elements typically come from different sources (see Fig. 6.7).

These conglomerates are now set in various business contexts and are adapted to these contexts through repeated combination and compensation. The initial business contexts can be drawn, for instance, from the strategy or specified by the project sponsor. These contexts can then also be tinkered with to arrive at a context in which the building block conglomerate exhibits its greatest value.

The entire process of adjusting building blocks and contexts is repeated until an optimal constellation results. A specific plan for further procedures is then determined on the basis of the result. Suboptimal results are stored in the warehouse. It is safe to assume that the number of business-optimized combo building blocks will be far fewer than the number of options available at the beginning of the process. The aim here is a reduction by a factor of around five.

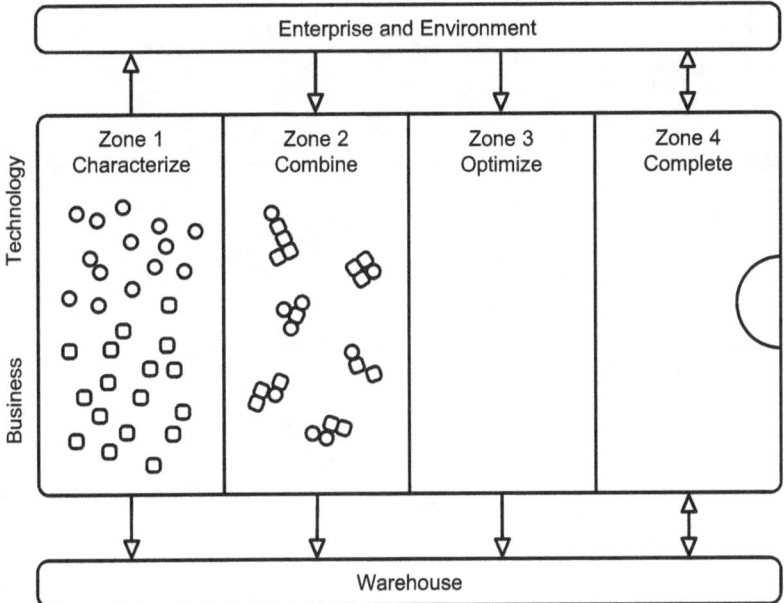

Fig. 6.7 Zone 2: combine

▶ Building block conglomerates are now set in various business contexts and optimally harmonized to the contexts via a process of combining and compensating.

New combinations, building block elements and building block fragments that can no longer be used are also stored in the warehouse as material that may be needed in the future. The new contexts that arise in the course of combining components are also stored in the warehouse or transferred to the strategy process.

Zone 3: Optimize

In the third zone, the number of building block conglomerates is again reduced, this time by a factor of around two. Possible future scenarios are worked out to provide a basis for arriving at a shorter list of prospective candidates. The selected conglomerates are then optimized to match the scenarios. The building block conglomerates that have now been assigned to scenarios are again adjusted and combined to create building block conglomerates with the highest possible commercial value. Variations that are not pursued are stored in the warehouse for possible future use. We remain in this phase for as long as it takes to arrive at around five or six highly attractive combo building blocks, including their accompanying contexts (see Fig. 6.8).

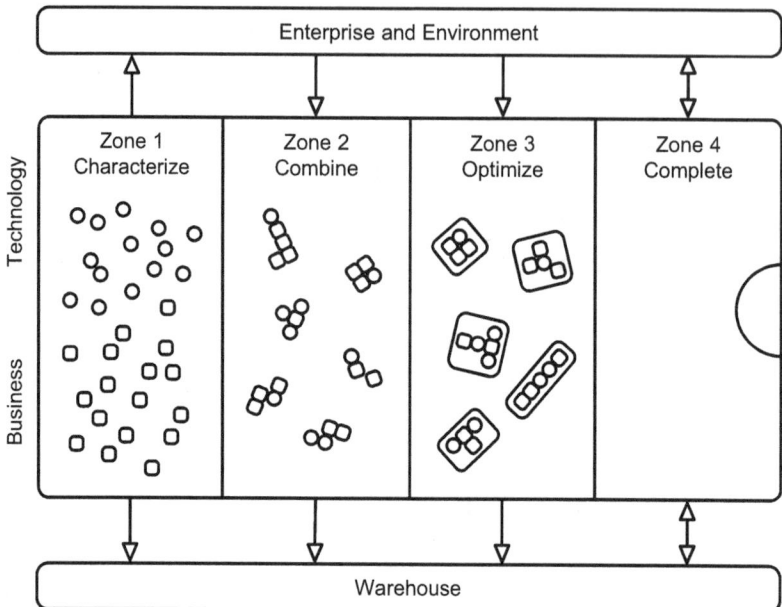

Fig. 6.8 Zone 3: optimize

It is altogether possible at this juncture that certain aspects of the remaining conglomerates are still somewhat utopian. This means that there remain aspects in the process that are still in need of clarification. This is accomplished in the fourth zone.

Zone 4: Complete

In this fourth zone, an attempt is made to clarify any missing or unresolved aspects of the highly attractive combo building blocks that have been identified. This involves identifying the questionable aspects and actively searching in all of the available sources for suitable solutions. Important sources here include the Internet and our own warehouse (see Fig. 6.9).

We can naturally not expect to bring our search to a successful conclusion in all cases. We can therefore count on losing certain combo building blocks. Such incomplete building block conglomerates are also stored in the warehouse and not simply thrown away! They may serve some purpose in another context, or we may find a missing part or solve a remaining puzzle.

Figure 6.10 offers a summary of our warehousing activities across the exploration zones.

It is very important for the exploration team to stay in close contact with the enterprise management and others in the enterprise environment throughout the

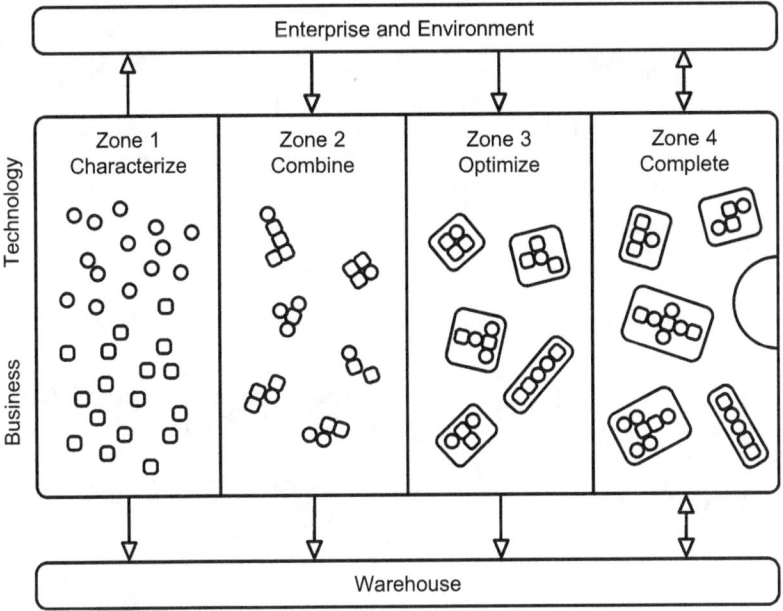

Fig. 6.9   Zone 4: complete

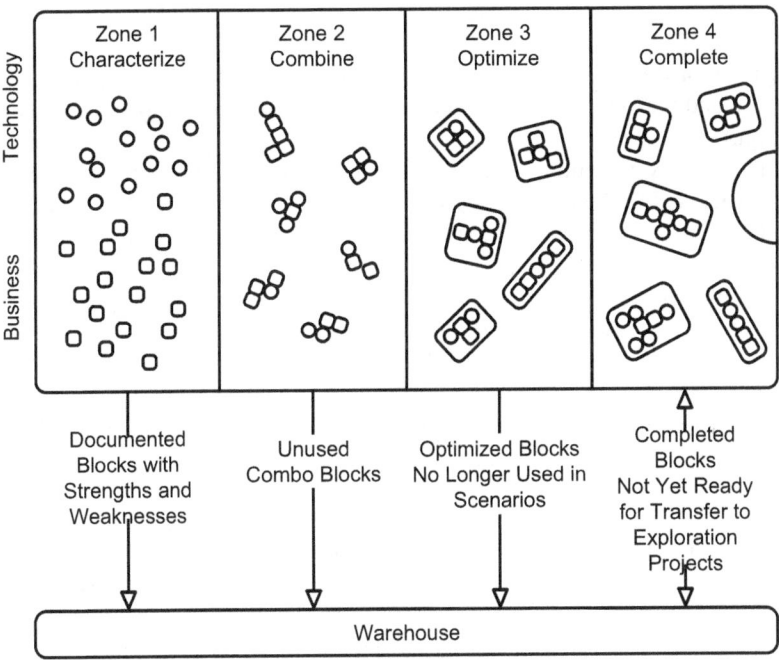

Fig. 6.10   A warehouse for intermediate results generated in the exploration arena

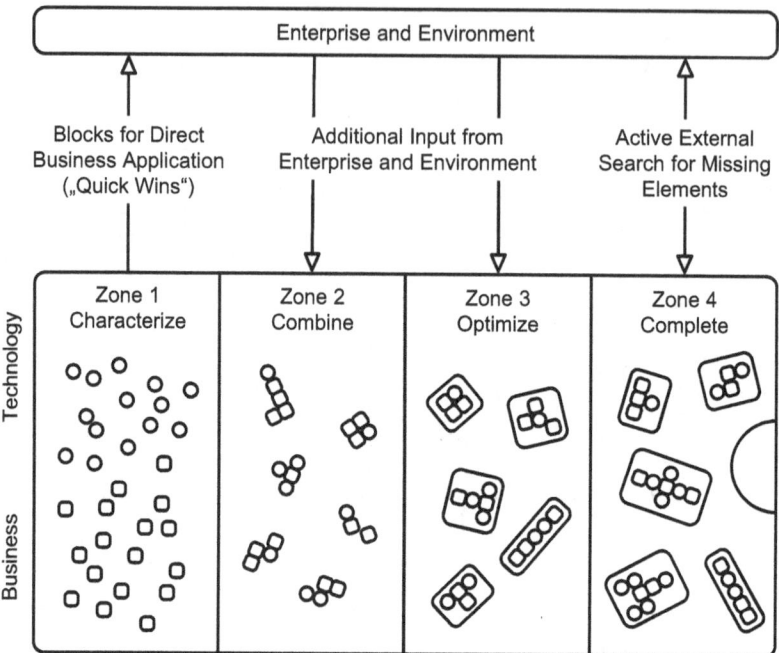

Fig. 6.11 Enterprise and environment

refinement process. Figure 6.11 illustrates the interaction that takes place between the team and the enterprise and others in the enterprise environment in the context of the different zones.

It is generally an option during the refinement process in the exploration arena to switch back and forth at will between the individual zones from 1 to 4. The process is indeed highly iterative and non-linear. The exploration team itself determines whether or not it is necessary to switch to other zones. It is therefore possible for a combo building block to circulate back and forth in the arena for a long time. There are no objections to be raised here so long as the combo building block continues to appreciate in value.

▶   The refinement process is highly iterative and non-linear. A combo
       building block may be shifted back and forth between the various
       zones of the exploration arena.

The proposed refinement process enables one to maximize the value of combo building blocks. Instead of simply evaluating the building blocks, these blocks are combined with others or parts of others to form conglomerates that then provide a basis for the development of mature business opportunities. Once the business case has been drafted in the decision-support subphase, the conditions for a holistic commercial evaluation will have been met for the first time.

▶   Building blocks are not simply evaluated, they are combined with others or parts of others to form conglomerates that are then refined into mature business opportunities.

The Development of a Business Opportunity in the Exploration Arena

Figure 6.12 offers an illustration of the refinement process for an individual business opportunity, and thereby helps to again clarify how the exploration arena works.

Technological and commercial building blocks in the first zone are joined together to form an initial building block conglomerate. A final business option then emerges from processes that involve combining and optimizing elements and identifying an optimal business context. The development across the zones of the playing field, however, is not linear. Reverse forays into zones that have already

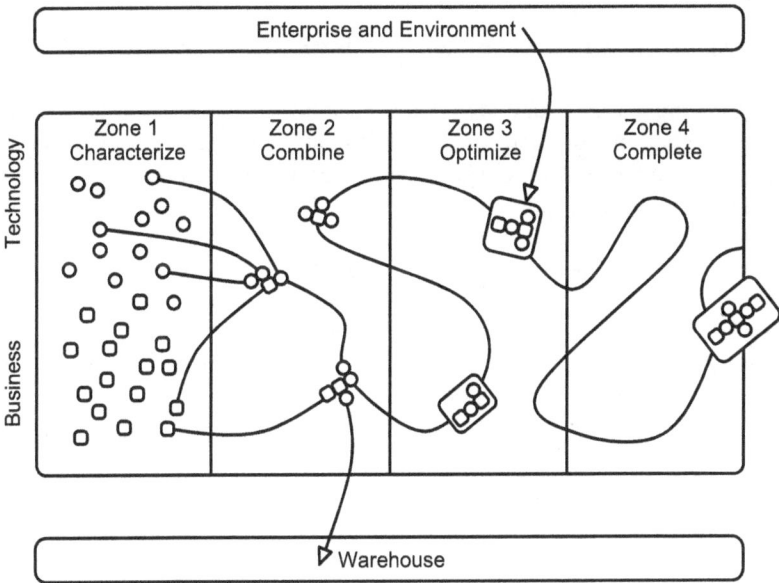

Fig. 6.12 The development of a business opportunity in the exploration arena

been crossed are also possible. In Zone 2, we see that a technology building block is stored in the warehouse after a determination has been made that it is no longer needed in connection with the current concept. In Zone 3, a business building block from the enterprise environment is added to further optimize the concept. The refinement process includes convergent and divergent phases and is, in general, an alternating process of analysis and synthesis.

Decision Support Through Exploration Projects

At the end of the exploration process in the exploration arena, we are left with five or six highly attractive business opportunities that are now positioned in the end zone. These business opportunities can now be submitted for further consideration. The decision as to whether a business opportunity is attractive enough to make it to the end zone is made by the exploration team in the context of working on conglomerates in Zone 4 of the exploration arena. There is no one else at this point who knows the business opportunity better or is in a better position to judge than the exploration team. The team therefore determines when a business opportunity is mature enough to leave the exploration arena.

▶   The exploration team determines when a business opportunity is
       mature enough to leave the exploration arena.

These proposals may naturally exhibit various degrees of maturity:

• Very mature business opportunities are submitted to the management with a recommendation for immediate implementation. Further development or processing in the context of an exploration project is not necessary.
• Less mature business opportunities are in need of additional exploratory work. In the case of technical aspects, the work may involve the development of prototypes or demonstrators. In the case of business aspects, the work may involve the further development of the business model (e.g. as outlined by Osterwalder and Pigneur 2010) and the definition of the minimum viable product or MVP (e.g. as outlined by Ries 2011, p. 33 f.). This phase is completed in the decision-preparation process in the form of an exploration project.

Example of an Exploration Project

The second exploration subprocess proceeds in a similarly dynamic and iterative manner. This subprocess centers on consolidating and verifying a business opportunity before turning it into a business case that is capable of promotion and evaluation. As in the case of the first subprocess (refinement) in the exploration arena, the exploration project is also to proceed in a holistic manner that ensures the simultaneous development of both technical and commercial aspects.

Figure 6.13 illustrates the setup for the subprocess. The upper half of the representation of the project includes the technological aspects (technologies, products, systems, etc.) and the lower half includes the business aspects (business contexts, markets, business models, customers, etc.). Depending on the current position of a given business opportunity on the vertical axis, the focus of the exploration will be on technological or business aspects.

The aim of the parallel development of the technological and commercial aspects is to significantly minimize both the technological and business risk in the exploration phase (see Fig. 6.14). Many enterprises begin by focusing only on the technological risks before only later in the development process moving on to an assessment of the commercial risks, or neglecting to do so altogether. This can lead as a consequence to big surprises and failure in the context of a product's market introduction.

The Exploration Project Portfolio

In what follows, let us again have a look at the decision-support subprocess. As described above, the process takes place in the framework of an exploration project. In order to increase the likelihood of success, it is generally advisable to have at least two to three projects running simultaneously. This number may naturally be

Fig. 6.13 Example of an exploration project

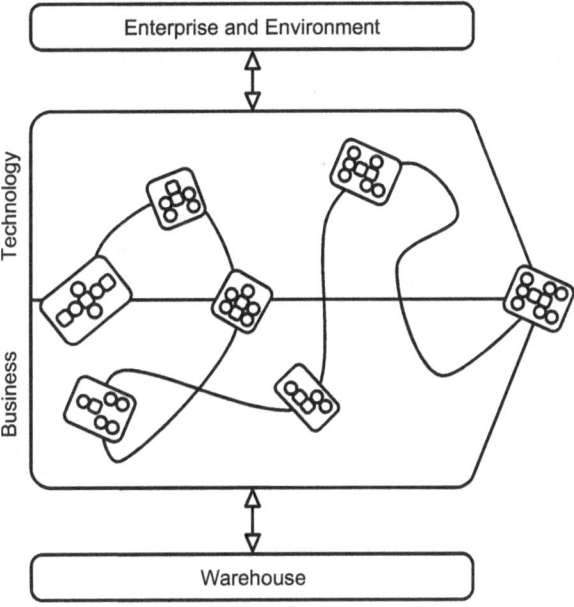

Fig. 6.14 Risk minimization

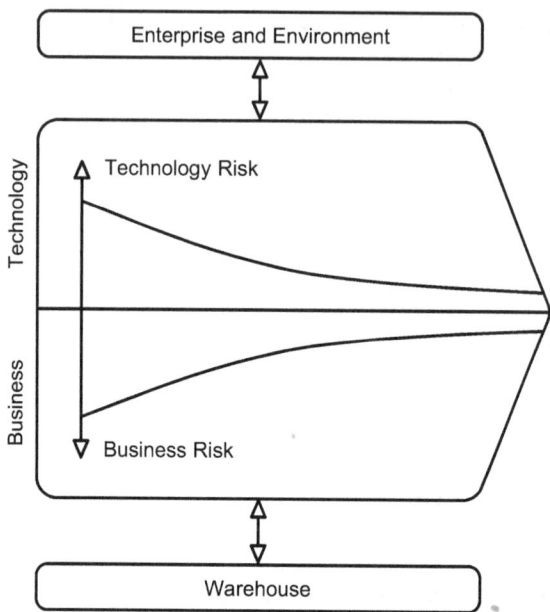

significantly larger in the case of large enterprises. We can then compile and holistically manage the parallel projects in a project portfolio.

Our experience shows that many enterprises, especially smaller ones, maintain many small development projects, or incremental innovations, alongside one major innovation project that often harbors considerable risk. This means that the long-term fortune of such enterprises may depend on the success of a single project.

It is therefore advisable to keep a portfolio of all of ones innovation projects. Figure 6.15 shows an exploration project portfolio containing four projects. Figure 6.16 shows a composite view of the exploration arena and exploration project portfolio.

The identified concept building blocks are entered in the exploration arena and refined in the four steps: characterize, combine, optimize and complete to form business opportunities. These business opportunities are then developed in the framework of exploration projects until they are ready for consideration in a decision-making process. It is in this project phase, i.e. when the business case is developed, that a commercial value can be assigned for the first time.

Those involved in the exploration phase maintain continuous contact with the enterprise management and the outside world throughout the entire exploration phase. It is especially important to search external sources for any missing information or concept building blocks. The outside world is also the source of any market information and of the view from the customer's perspective that are so important for the ultimate success of innovative products.

This function can be reinforced by availing oneself of the services of external exploration specialists. Such specialists can accompany the exploration activities in

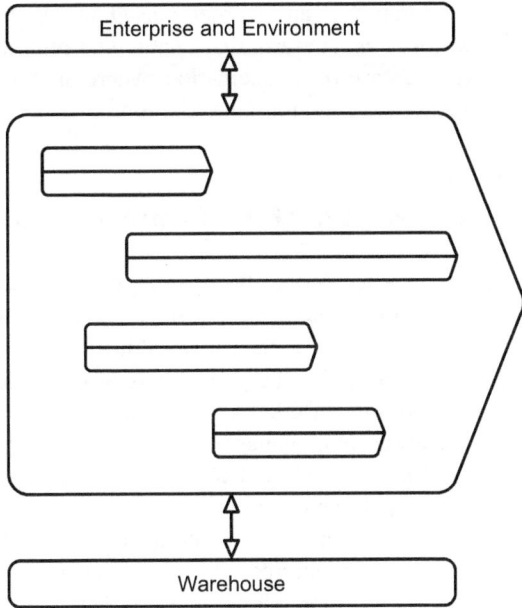

Fig. 6.15 Exploration project portfolio

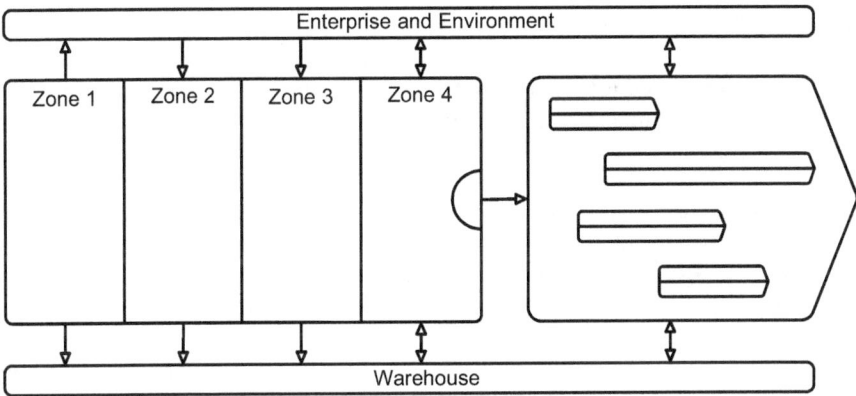

Fig. 6.16 Arena and portfolio

the capacity of a moderator or provide technical or business expertise. They can also provide a view of the activities from a neutral perspective and help to make sure that the exploration team is not implicitly limited to the current business model of the enterprise. When selecting outside consultants, it is important to make sure that they are properly experienced and qualified in the field of exploration (which is rare).

Furthermore it is also important to make sure that the knowledge acquired in the process is developed within the enterprise. This is why it is generally a bad idea to simply outsource the exploration to an external consultant. On the contrary, such consultants are only to serve as catalysts that enable and reinforce internal processes. It is a little like the case of a vaccination, where an external stimulus, a foreign body, is used to stimulate the immune reaction.

6.4   The Internal Structure of the Central Innovation Unit

Equipped with what we have learned so far, we can now proceed to the task of giving a structure to the internal organization of the central innovation unit.

We need to secure a different type of enterprise culture for the central innovation unit. We also need a different kind of human-resources and project management that is able to take account of this different culture and value system. Furthermore, the unit will need its own key account management, its own marketing team with an internally visible and independent communications department and its own brand appearance. In light of these additional features, one is unable to avoid the image of an enterprise within an enterprise. Indeed, the enterprise's central innovation unit is to be managed like an independent enterprise! While this naturally runs counter to the standard principles of organization, all of these privileges are indispensable for the proper functioning of the central innovation unit. This will be difficult for many within the enterprise to accept. Envy may begin to play a role and pressure may mount from time to time to close the innovation unit. Such pressure will further widen the innovation gap.

▶   The central innovation unit is to be managed like an independent enterprise. Ultimately, it will represent an enterprise within the enterprise.

We can now turn to a discussion of the central innovation unit's structure. Given that it represents an enterprise within an enterprise with its own playbook, it will naturally also have to have its own management with the usual management functions, including, in particular, its own department of human resources and its own finance department to operate according to a notion of budgeting that differs from that of the business divisions. We have also seen that an independent key account management and marketing department are necessary to secure professional interaction with the rest of the enterprise, which may well come across as foreign from the perspective of those in the innovation unit. And last but not least, the unit is to have an early warning system and an exploration process for the refinement of proposals in the exploration arena and the development of business opportunities in the context of exploration projects.

Our experience also shows that it makes sense to have the innovation unit perform its exploration activities in the form of projects. The packaging of activities in the form of projects gives undertakings that are generally difficult to plan a clear outward demarcation without limiting the crucial creative license within the project. The organization of exploration activities in the form of projects limits the risks (especially the financial risks) and makes them more manageable. It also permits one to more flexibly assign employees to specific projects. This is an important advantage when it comes to the deployment of the less manageable and largely self-motivated NT personality types. This mechanism allows NT-oriented empoyees to get deployed primarily in those areas in which they want to work.

It would even be feasible to have the employees apply for participation in projects. As result we get as a main organizational element sort of project factory in which work is available in project form. This means that the exploration employees are organized across organizational units that are ordered according to skill areas. From these units, they can then be dynamically assigned to projects as needed and in accordance with their skills, thereby creating a project factory of the type of a project-skills matrix.

Figure 6.17 illustrates the structure of the central innovation unit.

We have now given the central innovation unit a clear structure, including a project factory for exploration projects in the form of a project-skills matrix, a key account management with an integrated marketing unit, an early warning system and an independent management.

The organizational model can now be assigned to the various parts of the innovation process.

The early warning projects in the innovation process are represented on the early-warning-system subprocess. The project factory (with the exploration arena and the exploration projects) can also be assigned directly to the exploration phase

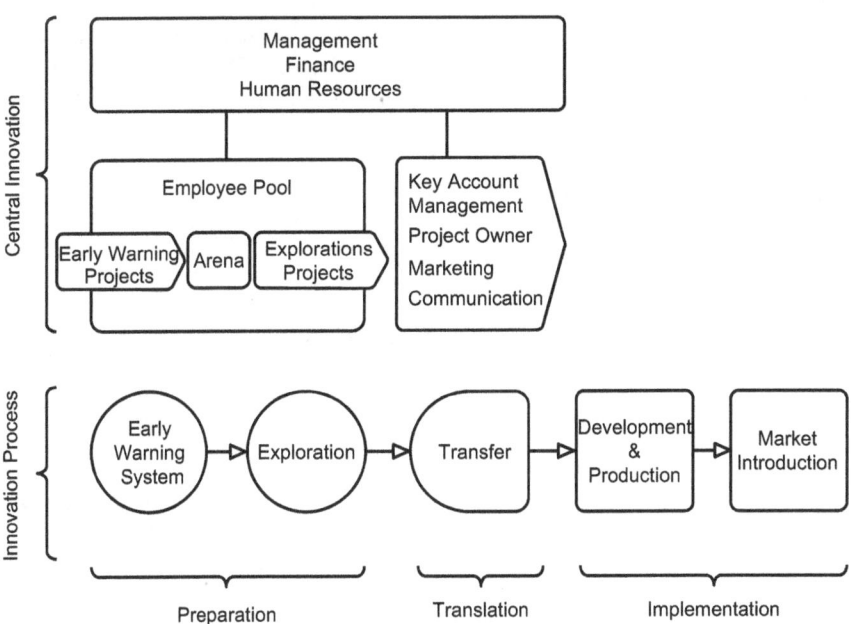

Fig. 6.17 Structure of the central innovation unit

in the innovation process. The key account management and its marketing department represent the transfer process. The central innovation unit is not responsible for the development and market-introduction process phases. The respective business divisions are responsible for these phases. The other functions of the central innovation unit (management, finance and human resources) form the overhead.

▶   The central innovation unit is not responsible for the development and
      market-introduction process phases. The respective business divisions
      are responsible for these phases.

The early warning system: As we suggested in Sect. 5.2, the purpose of the early warning system is to notify the enterprise of new developments and opportunities and to supply it with concept building blocks for the exploration arena. As shown in Fig. 6.18, we have used the image of a radar screen surrounding the enterprise to represent the early warning system.

Sources of such new opportunities and concept building blocks include the enterprise's environment and the enterprise's own employees. In particular, the early warning system is to perform the following two tasks:

1. To act as a sensory organ of the enterprise by systematically detecting and properly interpreting new developments in the relevant enterprise environment.
2. To encourage the generation of ideas by employees and to carry out an initial processing of these ideas, i.e. to tap into the enterprise's creative potential and to manage ideas.

Owing to the special nature of these tasks, it is clear that we will also need employees with special skills to perform them properly. We essentially need NT personality types with pronounced extraversion. Given that we will ultimately

Fig. 6.18 The early warning
system

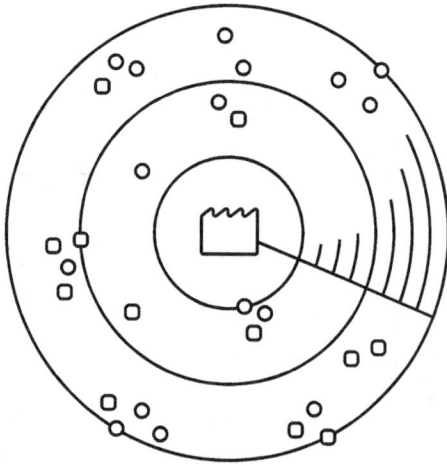

employ only a few individuals to perform these tasks, we are well advised to be very selective and to test the suitability of candidates in practice. Giving the early warning system the form of a project will again allow us to benefit from an optimal mix of a stable structure and internal flexibility. Employees who are less good at performing early-warning-system tasks can be easily reassigned to other tasks without losing face, something that is crucial in an NT-dominated organization because of the potential to generate fear in the organization through losing face. The best approach therefore is to secure the functioning of the early warning system in project form via our project factory.

▶ The functioning of the early warning system is secured in project form. This ensures an optimal mix of a stable structure and internal flexibility.

6.5 From Exploration to Innovation

Now that we have had a closer look at the exploration phase, we can return to our main pursuit, namely, innovation or the successful introduction of a novel product or process application to the market. The question now arises as to how we can best exploit the business opportunities that have been developed in the exploration process.

One solution is obvious. We can transfer our results to the relevant business division in the enterprise. While this is indeed the preferred approach of most enterprises to introducing new products to the market, it is not without its pitfalls. As we have suggested, the exploration process should ideally take place within a structure that enjoys a central position within the enterprise. The results of the exploration process will therefore have to be transferred from the central innovation unit to a business division. In Sect. 6.4, we indicated that this transfer procedure can only succeed if the relevant business divisions agree to the transfer. And this gives rise to the question as to what we should do if none of the business divisions within the enterprise are interested in the product's further development. There are many reasons why business divisions might not be interested:

• The division is already experiencing work overload.
• The division has other priorities.
• The individuals who would best be capable of handling the matter are not currently available.
• There is no way the budget would cover the expenses.
• Division representatives do not see any promise in the proposal.
• The proposal is seen as a threat to existing business.
• The incentives are not clear, with managers being rewarded for achieving other goals.
• The not-invented-here attitude gains the upper hand.

One option is to simply abandon the project. This is a typical response in real-world enterprises and is also a reason why many innovation projects end in failure.

What then are the available other ways of bringing innovation projects to a successful conclusion? One way is to create a new business division for the innovation project, either within the enterprise or outside of the enterprise in the form of a new startup enterprise. Another alternative would be to look beyond the enterprise to another enterprise that might be interested in buying the innovation project after the exploration phase. Or one could combine these latter two approaches by entering into a joint venture with a partner. As shown in Fig. 6.19, there are four ways of attempting at this juncture to exploit a prospective innovation:

- Setting up a new business division
- Founding a new enterprise
- Selling the innovation project
- Pursuing the innovation project with a partner in the framework of a joint venture

If we would like to be able to pursue any of these four ways of exploiting prospective innovations, then provision will have to be made within the enterprise for doing so.

For the first option of setting up a new business division, we don't need anything that goes beyond the existing functions. This decision can be made by the executive management or the supervisory board.

The situation is different if we're interested in selling the project. To do this, we'll have to protect the value generated in the exploration phase as much as possible. As the value involved is immaterial and cannot be locked up. It will therefore need to be kept secret. On the other hand, it will not be possible to conduct sales negotiations about secret contents. We therefore need a mechanism for protecting the immaterial or intellectual property involved. This is achieved by making use of the existing legal mechanisms for protecting intellectual property. The options center on patents, designs, brands, trademarks and other items subject to intellectual property rights (IPRs). In other words, we will need to provide for an

| New Business Division | Founding of New Enterprise | Sale of Project | Joint Venture |

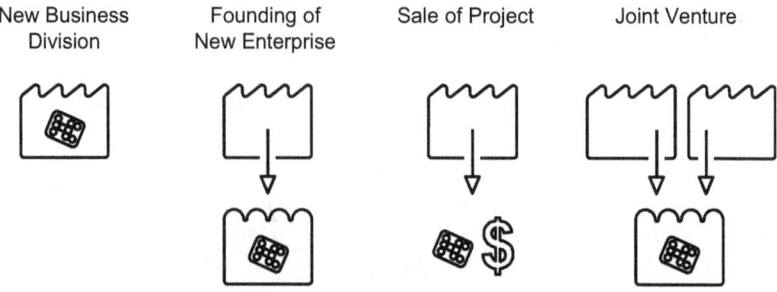

Fig. 6.19 Means of exploiting prospective innovations

additional role within the enterprise to manage intellectual property and to devise IPR strategies.

Let's now have a look at the option of founding a new enterprise. There are two prerequisites for this. First, one needs to know how to go about founding an appropriate enterprise. Second, one will need to secure ways of financing the new enterprise. For most enterprises, meeting these two prerequisites will be something new, a task that is usually handled by an additional enterprise role commonly referred to as venturing. Venturing specialists (e.g. venture-capital firms) usually insist that the new enterprise be in possession of intellectual property (e.g. patents). This case, too, will therefore require an IPR management.

In terms of the necessary enterprise functions, the last option of entering into a joint venture corresponds to the founding of a new enterprise, although with the addition of competence in the area of mergers and acquisitions. These are usually already available in larger enterprises.

For the Innovative Enterprise, this means the introduction of two new functions: IPR management and venturing. Both will naturally need to work closely with the exploration team. Furthermore, given that both are of strategic significance, they will need to interact closely with the strategy unit. In terms of organizational structure, they are to be positioned between the strategy and central innovation units, although it would also be possible place them within the above-mentioned organizational units. What is far more important than their placement in the organizational structure, however, is to secure ample opportunity for informal exchange between the employees of the various units. This can be best achieved by the sort of proximity that would allow the employees in the mentioned units to take coffee breaks together.

Figure 6.20 offers a view of the now complete organizational structure of the innovative enterprise.

6.6 The Rules of Exploration

The central innovation unit needs its own enterprise culture. For this reason, it represents a kind of enterprise within an enterprise, and this requires a number of additional independent features such as a separate marketing department.

The important factor here is that we need NT personality types for successful exploration. Such innovators tend to react differently to managerial intervention than the typical employee in a business division. The innovation unit therefore needs a management approach that is different from that of a typical business division within the enterprise. For instance, intrinsic motivation can be expected to play a far more significant role in the innovation unit.

How then are we to find the necessary NT personality types? This can essentially only be done with the use of a personality test. It is advisable to look for individuals with a mind of their own, individuals who may not be entirely assimilated and who distinguish themselves in terms of their ideas. NT personality types often resist

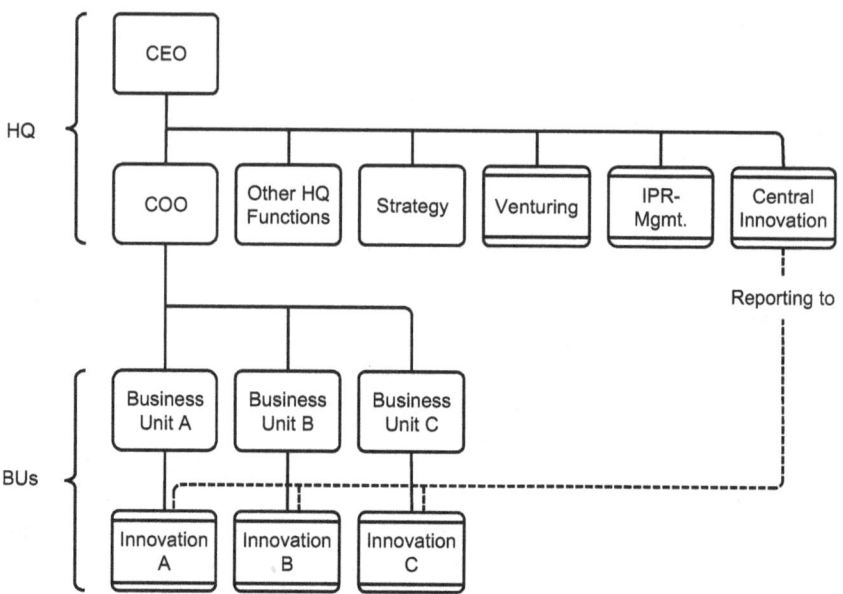

Fig. 6.20 The complete organizational structure of the innovative enterprise

detailed instructions and tend to call everything into question. They also typically have problems when it comes to self-organization and sticking to deadlines.

Exploration activities are to take place in the context of projects. The necessary organizational form for this is the project-skills matrix.

The exploration process is highly iterative and includes tasks that demand a high degree of creativity. While the process is difficult to plan in a classic sense, it can be managed in a systematic manner so long as the focus of ones approach is more on content than deadlines.

As indicated above, the transfer process is also handled by the innovation unit. The key individuals for this process are the key account managers in their role of transfer moderators. Here, too, it is important to give special consideration to employee assignment or recruitment policies. Suitable individuals for this role are typically well-assimilated in group contexts. These transfer moderators should also possess excellent communication skills, a capacity to accommodate others and a sense of diplomacy. Furthermore, they should be competent, reliable and meticulous, while at the same time showing a capacity to respond flexibly to changing situations.

6.7   The Case of the Swiss Postal Service

The Swiss Postal Service (Die Schweizerische Post AG) offers an example of how the application of the principles outlined in the present book can lead to a considerably enhanced capacity for innovation.

The Swiss Postal Service, a major corporation successfully active in multiple business sectors, introduced in the year 2012 an exploration initiative lead by Robert Käppeli. With the strength of its traditional business and solid enterprise structures as a basis, the Swiss Postal Service was interested in pursuing new business opportunities. The corporation's strengths include a comprehensive distribution network and high rates of acceptance and trust among Swiss citizens.

For years, the Swiss Postal Service has placed great value in innovation and has succeeded at introducing a number of highly regarded innovations. For instance, its web-stamp, which customers are allowed to design themselves, has been very popular. Despite an existing innovation unit and well-functioning innovation processes, the Swiss Postal Service was interested in improving and becoming even more successful. After commissioning InoBooster, an enterprise operated by the authors Daniel Huber and Martin Steinmann, the principles outlined in the present book were implemented. The first step was the creation of a new source of ideas. All of the company's 700 trainees at work in the final year of their apprenticeships were invited to participate in ideation workshops. For its part, the management was very curious about what these digital natives might come up with. And indeed, the project led to the generation and documentation of more than 100 ideas, many of which were quite remarkable. However, most of the ideas were essentially no different than ideas that had already been considered within the enterprise, a result that confirmed what we had observed over the course of many years. While the management was somewhat disappointed, it was clear that it would scarcely have been possible in the limited amount of time allocated to develop really unusual and striking ideas. As expected, the ideas were primarily centered on ways of improving the typical day-to-day life of the trainees. However, the result also showed that the enterprise's existing efforts were not entirely misguided.

The focus of the second step was then to concentrate the ideas with the help of InoBooster and to submit a number of proposals for implementation, as well as to generate additional, truly novel and holistic approaches from the extensive raw material. To do this, a total of four exploration workshops were organized. Despite the seemingly unspectacular input, this second step led to an array of promising results. Many of the original ideas had been enhanced. Moreover, more than 30 "Quick Wins", i.e. immediately implementable improvements, were identified and around 30 business ideas were submitted for further consideration and processing to various organizational units within the enterprise. A total of four new business fields were designated as strategic opportunities and were introduced to the enterprise strategy process in the form of initiatives. This final step was facilitated by the drafting of a small brochure to illustrate the results in an appealing and intuitive form and to thereby present them in an effective manner to the management.

Conclusion: The exploration process in this case proved to be valuable. It proved possible to fill the necessary roles with the right types of individuals and to validate and confirm the effectiveness of the new approach. The upcoming challenge is to widen the scope of the exploration process within the enterprise.

The application of the principles of exploration in the Swiss Postal Service turned out to be very successful and warrants emulation. The effectiveness of applying the principles outlined in the present book was confirmed. The playful exploration led to an improvement in the quality of the enterprise's approach to innovation.

References

Blank, S., & Dorf, B. (2012). *The startup owner's manual*. Pescadero, CA: K&S Ranch.
Johnson, M. (2010). *Seizing the white space*. Boston: Harvard Business Press.
Osterwalder, A., & Pigneur, Y. (2010). *Business model generation*. New York: Wiley.
Ries, E. (2011). *The Lean startup*. New York: Crown Business.
Taleb, N. (2013). *Kleines Handbuch für den Umgang mit Unwissen, German Issue*. Munich: Albrecht Knaus Verlag.

Managing Innovators

7

Throughout our discussion so far, we have outlined the organizational principles that need to be applied to give enterprises a capacity for innovation. In doing so, we have identified a number of situations that are destined to lead to power struggles and conflicts of interest. We have also shown how to circumnavigate these situations via a suitable organizational structure.

We now revisit the problems relating to communication and work culture that we addressed in Chap. 3. We have seen, for instance, how certain character traits that are regarded as virtuous in the preparation phase of the innovation process become vicious when applied to the implementation phase, and how other character traits that are regarded as virtuous in the implementation phase become vicious when applied to the preparation phase.

This presents us with a problem that cannot be solved by organizational means alone. As we will see, additional adjustments to our management principles will be necessary on account of the differences in the respective enterprise cultures of the different personality types.[1]

▶   The problems that arise in connection with the management of innovators cannot be solved with organizational means alone.

In Chap. 3, we identified the innovation gap and learned where and why there are basic communication problems between innovators in the early phase of the

---

All illustrations are published with the kind permission of © Heiner Kaufmann, Daniel Huber, and Martin Steinmann. All Rights Reserved.

[1] It warrants pointing out in this regard that not all employees are suitable for work in the two different phases. In addition to the relative suitability of different personality types, there is also the matter of different skill sets.

© Springer International Publishing AG 2017   121
D. Huber et al., *Bridging the Innovation Gap*, Management for Professionals,
DOI 10.1007/978-3-319-55498-3_7

innovation process and employees in the rest of the enterprise. In this regard, we formulated the following proposition:

> ▶ *Proposition 2: There is a heretofore unexamined innovation gap between the early phases of the innovation process (i.e. as represented by the funnel model) known as the ideation and evaluation-and-selection phases and the later implementation phases known as the development-and-production and market introduction phases.*

A look at the innovation process shows that the innovation gap appears between the exploration subprocess and the development subprocess. The source of the gap can be explained as follows: The early-warning-system and exploration subprocesses address the matter of *what* we are supposed to do. It is therefore not clear at the outset what the result of the corresponding subprocess will be. In Chap. 3, we suggested that NT personality types are more likely to succeed when it comes to approaching such open-ended tasks.

▶   NT personality types are better at successfully addressing the matter of *what* we are supposed to do.

In contrast, the development and market introduction phases involve determining the most efficient way of arriving at a clearly defined goal. And our suggestion here is that SJ personality types will be more likely to succeed when it comes to approaching such tasks whose preferred results have already been clearly defined. We have also learned that the two different personality types have very different ways of thinking. Given this background information, it stands to reason that the two groups will demonstrate very different employee behavior. And it is this employee behavior that we would now like to examine more carefully.

▶   SJ personality types are better at successfully approaching tasks whose preferred results are already clearly defined.

7.1   Management Principles for SJ Personality Types

Keirsey describes SJ personality types as follows[2]:

> *SJ type: SJ are the big organizers, the "Guardians". SJs claim: "I do it right". They can be described as the pillars of the established system; they are conservative, orderly, reliable*

---

[2]The older description used here from Keirsey.com from the year 2010 (http://www.keirsey.com/handler.aspx?s=keirsey&f=fourtemps&tab=3&c=overview) appears to us to be clearer and therefore more appropriate for our present purposes.

and accountable. SJ typically have an analytic deductive style of thinking, a process that is linear. They look for the optimum and for efficiency. For them the process is essential. SJs like hierarchical organizations where they integrate themselves smoothly. They need to feel secure and for that reason do not like changes at all. In short SJ are ideal managers and they blossom and perform best in operational business tasks.

This description and our practical experience indicate that employees who are SJ-oriented are likely to distinguish themselves in terms of the following qualities: professional skill, attention to detail, precision (bordering on pedantry), punctuality, discipline (specifications are construed rigidly as rules) and reliability. SJ-oriented employees do not allow themselves to be distracted by environmental influences from the defined goal. They also place great value in rational analysis as a behavioral guide. Decisions are to be made in a rational and dispassionate manner. When it comes to management, SJ personality types are therefore easy to reach via rational incentives (e.g. bonuses for outstanding performance). Given that they regard emotions as more of an annoyance, their reactions to emotional incentives are less strong or do not involve a direct awareness.

What does this mean when it comes to day-to-day management? SJ-oriented employees can be optimally deployed when the goals are clear. They respond well to classic management signals including goals, specifications, monetary incentives and penalties. In short, they are ideal enterprise employees in nearly all contexts. The classic manuals on management are very effective in the case of SJ personality types.

The management situation is very different when it comes to the NT personality types we depend on for the success of the two early phases in the innovation process (early warning system and exploration).

7.2   Management Principles for NT Personality Types

We have suggested throughout the previous chapters that NT personality types tend to think in a manner that is altogether different from that of SJ personality types. It is therefore not surprising that they will respond more favorably to and be more productive in a different enterprise culture. This was the basis for our endorsement of an independent organizational unit for NT personality types. Given the apparent need of NT personality types to adhere to a different set of values, we suspect that it will be essential to apply a different set of management principles to lead them.

Keirsey describes NT personality types as follows[3]:

NT type: NTs on the other hand are the keepers of knowledge, the "Rationals". NTs claim: "I understand how it works". They can be described as knowledge-centered experts; they are creative, conceptual, factual and impersonal. NTs typically learn incessantly, without

---

[3]The older description used here from Keirsey.com from the year 2010 (http://www.keirsey.com/handler.aspx?s=keirsey&f=fourtemps&tab=3&c=overview) appears to us to be clearer and therefore more appropriate for our present purposes.

thinking to apply their knowledge: to know is everything. But if they do apply their knowledge, the result will be first class. They typically think in analogies and work deductively. They don't like hierarchies, but form networked organizations instead. NTs perceive changes as interesting experiments. In short NTs are the ideal researchers and they blossom and perform best in unknown territory.

This description and our practical experience indicate that employees who are NT-oriented are likely to distinguish themselves in term of the following qualities: professional skill, flexibility, strong will (bordering on stubbornness), specifications are construed flexibly as "wishes," sensitivity to the environment, creative interpretation of tasks given changes in the environment. NT-oriented employees make use of their emotions when looking for solutions. They attempt to feel their way into their tasks in a wholistic manner. They are aware of the fact that their tasks cannot be solved in a purely rational fashion. NT employees can therefore be managed via their emotions. In contrast, they often remain unfazed by purely rational incentives.

We also face the following further problem in the context of the two early phases of the innovation process. It's not possible to command people to have creative ideas. After all, what we need are people who are engaged in their work for its own sake. Here, we are advised to recall Antoine de Saint-Exupéry's remark about what it takes to build a ship:

If you want to build a ship, don't drum up people to collect wood and don't assign them tasks and work, but rather teach them to long for the endless immensity of the sea![4]

We will therefore have exactly the people we need during the early innovation phases if they perform their assignments out of a direct interest in doing so. Their motivation has to be as intrinsic as possible. Extrinsic incentives such as specifications, rewards and penalties will be less effective in such environments.

Indeed, NT-oriented employees do not respond well to the classic enterprise management signals. They are not really inspired by monetary incentives. To them, money is a mere matter of hygiene. It needs to be part of the picture to an appropriate extent, but is otherwise irrelevant. Reactions, often in the form of a resignation, come only when the money factor is way out of alignment. Goals and rules are generally regarded by NT personality types as non-binding wishes. Owing to their independent streaks, they primarily follow their own intrinsic goals. When threatened they may simply retire. They see threats of small penalties as petty and ridiculous. They tend to ignore them. Efforts to exert greater pressure are likely to be met by spite and resistance. If they perceive the system around them as threatening, they begin to act in subversive ways which, given their creativity,

---

[4]Frequently cited and presumably erroneous "quote" of Antoine de Saint-Exupéry. Saint-Exupéry's work *Wisdom of the Sands* is normally given as the source (sometimes erroneously *The Little Prince* where nothing of the sort comes to pass). The corresponding passage, however, is somewhat different: "When I communicate to my people the love of sailing so that each and every one senses a yearning in himself because a weight inside himself pulls him to the sea, then you will soon see how they look for different activities that promise them a thousand special properties." (Saint-Exupéry 1951, p. 285 in the German edition).

can pose a big risk to the enterprise. In short, NT-oriented employees cannot be managed at all using the classic methods! They essentially do what they want to do.

This behavior is naturally a nightmare for the average manager in the enterprise. This is why individuals with pronounced NT traits often wind up in conflicts on multiple fronts and are frequently removed from the enterprise after a short stay. While this is also an altogether acceptable approach within the business divisions of an enterprise, it doesn't apply to the central innovation unit where the constructive engagement of NT personality types plays such an important role! So how are we supposed to manage the central innovation unit?

7.3   Prerequisites for Managing Employees with NT Traits

The focus of management in such an environment is different from the focus of classic management. Given that employees with NT traits essentially cannot be told what to do, and more or less decide this for themselves, the management challenge consists of finding out what their personal dreams and goals are, and then bringing these into alignment with the enterprise goals and the wishes of customers. From the perspective of NT-oriented employees, the enterprise is called upon to bring their personal goals into alignment with those of the enterprise. And this is naturally the ultimate aim of all managers! People reach their maximum performance levels when their personal commitments match those of the larger enterprise.

► The management challenge consists of finding out the personal dreams and goals of the employees, and then bringing these into alignment with the enterprise goals and the wishes of customers.

So how do we manage the activities of employees with NT traits when we know full well that the usual incentives approaches don't work? The answer is that we manage their activities without the classic incentives system. Actually, we treat them the way everyone prefers to be treated, with social competence. And indeed, social competence is the crucial criterion for successful management in an environment where gifted explorers are free to consider what is possible.

▶   Social competence is the key to managing employees with NT traits.

The Task of Managing Innovators
The keys to socially competent management include:

- Leading by example, with the heart, the mind and the hand (Heinrich Pestalozzi), treating employees as social and emotional beings
- Assuming (and communicating) that all are doing their best
- Building trust by consistent behavior and transparent communication
- Developing ambition while looking for and appreciating (highlighting) the positive
- Communicating and justifying clear visions and goals
- Celebrating success
- Deciding on the basis of consensus and promising solutions and sticking to decisions
- Making as much information as possible available to all
- Permitting a healthy measure of participation
- Deploying and promoting employees properly
- Organizing and integrating employees in teams
- Working hand in hand
- Maintaining a sense of humor

7.4   Managing Employees with NT Traits

In addition to social competence on the part of the manager, the most important factor in managing employees with NT traits is to take account of the personal development of the employees. This involves the following three approaches: supporting the careers of employees, motivating the employees and reinforcing the self-confidence of employees. Given that all three hinge on the inner life of the employee, managers can only influence them indirectly, namely, by adjusting their own behavior in recognition of the character of the employee. Three approaches can be identified: showing appreciation for employees, being favorably disposed toward employees and demanding excellent performance on the part of employees. These represent areas in which managers can and must act.

As shown in the diagram in Fig. 7.1, the above-mentioned six approaches are heavily interdependent.

The diagram is to be interpreted as follows: when applied together with appreciation, demanding excellent performance of employees leads to motivation; showing an appreciation of and a favorable disposition towards employees leads to a reinforcement of their self-confidence; when applied together with a favorable disposition toward employees, demanding excellent performance will support the careers of employees. Each effect therefore depends on the fulfillment of two conditions. If

Fig. 7.1 The leadership
triangle

these conditions are not met, then the result will tend to be negative. Managers who demand excellent performance without showing a favorable disposition towards employees will fail to support them in their careers and instead lead to their burnout.

These management principles apply to all people. However, given that we have no recourse to the standard, incentives-based motivational systems in the case of employees with NT traits, these management principles take on a far greater significance.

While the usual incentives-based systems of motivation (e.g. involving monetary rewards) don't work in the case of NT-oriented employees, this does not mean that all incentives-based systems are destined to fail. The incentives need only to be adapted to NT personality types, i.e. they should be oriented towards the employees' favored areas of inquiry or their personal development. For instance, employees with NT traits can be motivated with new laboratory equipment, other workplace enhancements or trips to conventions for specialists. A few general rules for the management of employees with NT traits apply:

Rules for Managing Employees with NT Traits

- Employees must be intrinsically motivated by the content of their work. Taking account of this aspect is especially important in the context of recruiting new employees.
- While it is important to set challenging goals, failure to reach the goals should not be met with penalties. Nothing is won by encouraging risk-averse behavior. In the words of H. O. Rohner, "Experience the joy of the wind while sailing instead of the fear of the high waves."[5]

(continued)

---

[5]Hans O. Rohner, www.b4u.ch.

- Use soft development measures when responding to deficient performance. Explorers need to feel free to venture into unknown terrain. Everyone is a beginner in the world of exploration.
- Staff downsizing is to be avoided at all costs, as it will tend to generate fear and extinguish creative license.[6]
- Provide a good salary, but not a top salary. Otherwise, one may wind up attracting salary-minded individuals with no clear NT qualities.
- It is advisable to avoid monetary incentives and bonuses.
- Ensure a large degree of independence and extensive liberties when it comes to work subjects. NT personality types respond favorably to such freedoms.
- Provide employees with opportunities to gather information and to advance their knowledge such as participation in advanced training programs and visits to conventions at home and abroad.
- Provide opportunities for publications and other means of presenting results.
- Grant extensive freedoms when it comes to work hours and methods.
- Show a clear willingness to allocate adequate resources (i.e. within the means of the enterprise). By allocating too little, one runs the risk of obtaining no more than obvious results because the employees don't have enough time to develop original ideas.
- Avoid administrative pressure and detailed rules.
- Apply clear sanctions in response to the abuse of freedoms.
- Promote internal transparency (e.g. forums for the presentation and discussion of results among colleagues).

7.5   Managing Exploration Projects

In addition to taking a different approach to the management of the central innovation unit, it is also essential to take account of the different nature of exploration projects and to manage accordingly, i.e. on the basis of different rules. This applies equally to management practice within of the project as well as to the interaction of the project manager with his clients. This is because exploration projects are best carried out by employees with NT traits. And as we have suggested, employees with NT traits are best managed with a different set of rules. We are essentially up against a different culture and a different type of employee who can be expected to react to different incentives.

One other important difference is that exploration projects expressly do not involve the pursuit of a clearly defined goal. Lacking this, exploration projects are essentially not amenable to planning. The associated costs cannot be reliably

---

[6]The aim here is to preserve the intrinsic motivation of the employees. Staff cuts are poison.

Table 7.1 Rules for managing exploration projects

Business projects	Exploration projects
The focus of project management is efficiency	The focus of project management is substantive quality (effectiveness)
Mile stone reviews are used to control on-time and on-budget achievement of established goals	Regular reviews permit progress assessments
	Deadlines and budgets cannot be planned owing to the indeterminate nature of the goals
Project reviews must be carried out in a very focused manner. Calling the goals into question is to be avoided	Project reviews must take account of many different contexts. Questioning and varying goals is one of the main tasks
Adhering to the budget and the deadlines is critical	It is essentially not possible to set budgets and deadlines. While this makes the monitoring of such a secondary task, it is necessary project hygiene. It represents the only way of ensuring that resources are used responsibly
The project manager is a deadline task master, a pilot and sometimes a dictator	The project manager is the team coach
Project review meetings are of the type task distribution according to the project plan	Project review meetings are coached workshops with an emphasis on the intrinsic motivation of the project teams

estimated and milestones cannot be set. Deadline and budget overruns thereby take on a whole new meaning and need to be treated accordingly, i.e. as outlined in Table 7.1.

Reference

de Saint-Exupéry, A. (1951). English edition: *Wisdom of the sands*. New York: Harcourt, Brace and Company (Citadelle 1948). (German edition: *Die Stadt in der Wüste* (Dt. Ausgabe). Düsseldorf: Rauch).

Exploration and Strategy

8

8.1   What Is Strategy?

Let's first consider the term strategy. Gabler's dictionary of business terms defines the term strategy as follows[1]:

> *Strategy is the basic, long-term behavior of the enterprise and any relevant divisions thereof as this is manifested in the measures it implements with respect to its environment so as to achieve its long-term goals.*

Wikipedia says the following about the term (business) strategy[2]:

> *Business (or Strategic) management is the art, science, and craft of formulating, implementing and evaluating cross-functional decisions that will enable an organization to achieve its long-term objectives. It is the process of specifying the organization's mission, vision and objectives, developing policies and plans, often in terms of projects and programs, which are designed to achieve these objectives, and then allocating resources to implement the policies and plans, projects and programs. Strategic manage-ment seeks to coordinate and integrate the activities of the various functional areas of a business in order to achieve long-term organizational objectives. A balanced scorecard is often used to evaluate the overall performance of the business and its progress towards objectives.*

> *Strategic management is the highest level of managerial activity. Strategies are typically planned, crafted or guided by the Chief Executive Officer, approved or authorized by the Board of directors, and then implemented under the supervision of the organization's top*

All illustrations are published with the kind permission of © Heiner Kaufmann, Daniel Huber, and Martin Steinmann. All Rights Reserved.

[1]http://wirtschaftslexikon.gabler.de/Definition/strategie.html#definition, as referenced on June 27, 2013.
[2]https://en.wikibooks.org/wiki/Business_Strategy, as referenced on April 14, 2016.

© Springer International Publishing AG 2017   131
D. Huber et al., *Bridging the Innovation Gap*, Management for Professionals,
DOI 10.1007/978-3-319-55498-3_8

management team or senior executives. Strategic management provides overall direction to
the enterprise and is closely related to the field of Organization Studies.

It is not our intention to select a specific definition of the term strategy or go as
far as to propose a new definition of the term. After all, the primary aim of the
present book is to examine the subject of innovation and not strategy. However, we
would like to clarify the ways in which strategy and innovation influence one
another and use this as a basis for deriving the prerequisites enterprise strategies
will have to meet so as to ensure that the enterprise is also strategically prepared for
innovation. And when it comes to this objective, the definitions of strategy cited
above are sufficiently precise. In any case, the definitions cited above permit the
following statements:

- In the field of economics, there is apparently no uniform definition of the term
  strategy.
- Strategy is essentially concerned with the long-term planning of enterprise
  activities. This can be broken down further according to time horizons or other
  criteria.
- A strategy is apparently supposed to indicate how the enterprise would like to
  behave in the future.

These terminological statements suffice for our purposes. Let us now return to
our main concern. What prerequisites does an enterprise strategy have to meet to
ensure that the enterprise has a capacity for innovation? Here, it is important
to recall what innovation is: to successfully introduce something new and relevant
for the enterprise. We also know that innovation is crucial to the future of the
enterprise. Recall the words of Chesbrough, "Most innovations fail. And companies
that don't innovate die." In keeping with the above-mentioned definition of strategy, innovation is something that must be very strategic.[3] After all, it is a matter of
creating the future of the enterprise.

However, if we take a look at real-world enterprise strategies, we usually find
something like the following: a description of the enterprise's business segments, a
plan for the development of these segments, including internal and external changes
and their drivers, a strategic measures plan and the accompanying financial plan.
This status also essentially corresponds to the requirements and methods conveyed
in textbooks on the subject of strategy (e.g. Lombriser and Abplanalp 2005). The
majority of real-world enterprise strategies make no mention of the subject of
innovation, and if they do, then it is usually limited to generalizations (e.g. "We
aim to position ourselves as an innovative enterprise.")[4]

---

[3]The strategy of an enterprise essentially reflects its DNA or the values it seeks to embody.

[4]This often derives from a limited understanding of innovation as nothing more than new ideas. As
we have seen, this is an inadequate conception.

Fig. 8.1 Dual strategy
according to Derek Abell

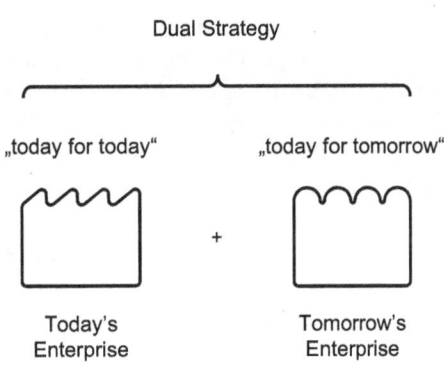

Dual Strategy

„today for today"   „today for tomorrow"

+

Today's   Tomorrow's
Enterprise   Enterprise

8.2   Two Strategies

There are, however, a number of strategy experts who have addressed the issue of innovation. We would like in particular to mention Derek Abell in this connection.[5] Derek Abell has introduced the concept of the dual strategy (e.g. Abell 1993 or Abell 1999), which is shown in Fig. 8.1. The idea here is that one strategy concerns today's business while a second strategy is devised to address the structure of future business segments. Abell refers to the first strategy as "planning for today" or the "today-for-today strategy." This strategy says what needs to be done today to optimally manage today's business. This today-for-today strategy essentially corresponds to the strategies one finds in most enterprises today.

Abell refers to his second strategy as "planning for tomorrow" or the "today-for-tomorrow strategy." This strategy says what needs to be done today to secure good business for tomorrow. This second today-for-tomorrow strategy has clear innovation character. In more recent publications, this second strategy is often referred to using the term innovation strategy. However, the innovation strategy

---

[5]*Derek F. Abell is the founding president of ESMT European School of Management and Technology and Professor Emeritus. His primary teaching and research interests are in strategic marketing, general management, leadership in technology-based industry and the broader responsibilities of the executive "beyond the bottom line." From 1981–2003, he was Professor of Strategy and Marketing at IMD, Lausanne, and parallel to that, from 1994–2003, Professor of Technology and Management at the two Swiss Federal Institutes of Technology (ETH Zurich and EPF Lausanne). He was Dean of IMEDE (the institutional predecessor of IMD) from 1981–1989, and from 1969–1981 a full-time faculty member at the Harvard Business School.*

He originally graduated as an aeronautical engineer from the University of Southampton. He has a Master's degree from the Sloan School of Management at MIT and his Doctorate from Harvard Business School. He has served as a consultant to governments in Eastern and Central Europe, as well as to many multinational corporations around the world. He has also served as a board member of a number of European-based public and private organizations. He has published five books and numerous articles. His most recent book is Managing with Dual Strategies: Mastering the Present, Preempting the Future, published by Free Press, New York.

Link: http://www.esmt.org/derek-f-abell, as referenced on June 27, 2013.

is typically understood to mean a functional substrategy that is subordinate to the enterprise strategy. This does not correspond to Abell's today-for-tomorrow strategy, which he sees as the second integral part of the dual enterprise strategy. If we consider the fact that the purpose of the innovation strategy is to secure the future of the enterprise,[6] then it becomes clear that Abell offers a more cogent description of the phenomenon.

▶   According to Derek Abell, enterprises need two strategies: a today-for-today strategy corresponding to the standard enterprise strategy and a today-for-tomorrow strategy corresponding to an innovation strategy.

8.3   The Innovation Strategy

In keeping with the today-for-tomorrow strategy, the innovation strategy encompasses all of the strategic considerations and measures that are necessary today to secure good business development tomorrow. There is surprisingly little about this subject in the literature. It is apparently not yet clear what an innovation strategy is expected to encompass.[7]

Let us therefore first have a look at the purpose of devising an innovation strategy for the enterprise. We know that the innovation strategy is supposed to enable innovation within the enterprise and to thereby help secure the survival of the enterprise in the future. The strategy is to specify the measures that will have to be implemented to ensure the enterprise's capacity for innovation. When we consider what this amounts to exactly, then we find much of what we have described in the present book. These measures can be summarized as follows:

• Analyze the enterprise strategy: Any innovation strategy must take account of the general enterprise strategy. Here, it doesn't matter whether or not a strategy has actually been formulated. To be able to draft an innovation strategy it is of course indispensable to have a clear understanding of the enterprise's current self concept (what the enterprise currently is), the concept of itself that it would like to establish in the future (what the enterprise wants to be in the future).

---

[6]The only thing we know about the future with certainty is that it will not be like the present. For this reason alone, we can safely assume that future enterprises will operate differently than they do today. Our experience shows that the longevity of business segments is limited. At some point, the demand for the corresponding products breaks away and the segments become obsolete. This is why innovative enterprises are the only ones that stand a chance of long-term survival.

[7]In one of the few books that address the issue of innovation strategy, Beat Birkenmeier and Harald Brodbeck describe the following five parameters of innovation: innovation strategy, innovation process, innovation instruments, innovation structures and innovations culture. However, Birkenmeier and Brodbeck also paint a rather diffuse picture of the innovation strategy (Birkenmeier and Brodbeck 2010).

Similarly, it should also be at least roughly clear where the barriers are on the path to the future. The aim of analyzing the enterprise strategy is to work out the strategic challenge faced by the enterprise.

If the enterprise strategy has not been documented, then it will be necessary to at least gather the relevant information from the behavior the enterprise has shown so far, as well as from interviews with the executive management.

- Analyze any existing innovation strategies: It is necessary to analyze and evaluate any existing innovation strategy.
- Set the innovation goals: It is essential to make clear where our innovation efforts are to lead us. Here, it is also important to make sure that the various innovation goals are mutually compatible and add up to a holistic vision. The target areas for exploration need to be established.
- Analyze the current state of our innovation efforts: We need to take account of what is already in place and available because that will define our starting point. In particular, we should be in the clear on what our current business model is.
- Establish or review the innovation process: What does our existing innovation process look like? What does our target innovation process look like? The model process outlined in Sect. 5.3 can be used as a guide in this connection (see Fig. 5.16).
- Establish or review the innovation organization: What does our existing innovation organization look like? What does our target innovation organization look like? The model organization outlined in Sect. 6.5 can be used as a guide in this connection (see Fig. 6.20).
- Establish or review the innovation budget: We have to know how much we want to invest in our innovation program and how much we can afford . . . and whether we can afford to refrain from investing in this area.
- Establish or review the enterprise's innovator management policy: What management principles are currently (de facto) applied? What principles would we like to apply? The management principles outlined in Chap. 7 can be used as a guide in this connection. In what ways do we need to adjust the existing incentives?

The crux of the matter is to create the conditions within the enterprise that will enable innovation to take hold and to flourish. So why have we refrained from mentioning innovation culture in this context? Innovation culture is not a subject that can be directly controlled. It is more a matter of a consequence of the favorable conditions we create (Zillner and Krusche 2012, p. 252). In addition to organizational structure, our management policies will be particularly important when it comes to establishing a favorable innovation culture.

▶   Innovation culture is not a subject that can be directly controlled. It is more a matter of a consequence of the favorable conditions we create.

8.4   The Relationship Between Exploration and Strategy

As we learned in Chap. 6, exploration is primarily a process of searching that leads to the refinement of ideas that have been submitted as input. This process is highly iterative and requires a large measure of creativity. It may turn out that the results, the new business opportunities, no longer have much in common with the originally submitted ideas. Just as in the case of any search, there is the searching part and the finding part. In other words, we sometimes find things that are different from what we set out to find. And these other things can be very attractive.

We can describe the innovation process as follows. In the early innovation phases, early warning system and exploration, we explore in those regions that are specified in the innovation strategy. We should therefore have good odds of finding new business opportunities. However, we cannot rule out the possibility of finding attractive business opportunities that are beyond the scope of our innovation strategy or even beyond the scope of our enterprise strategy, opportunities which we actually did not even set out to find. Indeed, exactly this happens quite often. If the results are attractive enough, then it may well be worth it to revise our prior strategies! This means that while we allow our strategy to determine our innovation activities, it may turn out that our innovation activities wind up determining our strategy. As illustrated in Fig. 8.2, exploration is a kind of activity that also provides input for strategy development.

This is why it is important to house our strategy unit and our central innovation unit in the immediate vicinity of one another and to provide for ample occasions for the informal exchange of ideas (e.g. a shared lounge).

▶ In the early innovation phases, early warning system and exploration, we explore in those regions that are specified in the innovation strategy. However, we cannot rule out the possibility of finding attractive business opportunities that are beyond the scope of our innovation strategy or even beyond the scope of our enterprise strategy, opportunities which we actually did not even set out to find. Indeed, exactly this happens quite often. If the results are attractive enough, then it may well be worth it to revise our prior strategies! This means that while we allow our strategy to determine our innovation activities, it may turn out that our innovation activities wind up determining our strategy. Exploration is a kind of activity that also provides input for strategy development.

Fig. 8.2 Interaction between strategy and innovation

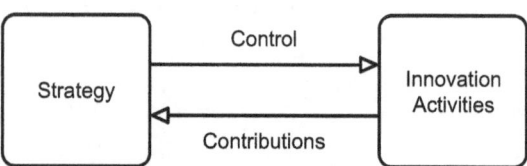

8.5   Intrapreneurs

In his famous book *The Innovator's Dilemma*, Clayton Christensen describes what he refers to as disruptive innovations (Christensen 1997). In contrast to sustaining innovations, disruptive innovations wind up changing the market (Christensen 1997, p. xvii). Christensen then suggests that disruptive innovation always need to be developed in the framework of a separate enterprise if they are to succeed (Christensen 1997, p. 113, 117).

Other authors address the subject of business models. Warranting special mention in this regard are Henry Chesbrough (2003, p. 63), Alexander Osterwalder and Yves Pigneur (2010) and Mark Johnson (2010). Here, too, the prevailing opinion seems to be that each independent business model will require its own enterprise or its own business division. For instance, Johnson writes, "... *it's nearly impossible for a business unit to adopt and operate more than one business model at a time.*" (Johnson 2010, p. 167) Elsewhere, Johnson describes a discussion between Clayton Christensen and Andy Grove, the former CEO of the Intel Corporation, from the year 1999 in which Grove asserts that, "... *disruptive threats (come) inherently not from new technology, but from new business models.*" (Johnson 2010, p. 191) Johnson goes on to describe the following consensus: "*Since then, we have realized that disruptive innovation and business model innovation were opposite sides of the same coin.*" (Johnson 2010, p. 191) This makes clear that disruptive innovations are essentially business model innovations,[8] which is why Christensen, too, calls for separate business units for disruptive innovations.

▶   Disruptive innovations are essentially business model innovations.

One can arrive at the same result via other considerations. For instance, we know that creating a business model is a very complex process. There are usually no individuals within an enterprise who have a complete grasp of the existing business model. This is at least the case with older business models that have since been allowed to grow organically. Once a business model has been implemented, it is not long before the perception of managers and employees is limited to only a part of the model. While this is altogether desirable because it permits a useful reduction in enterprise complexity (cf. Zillner and Krusche 2012), it leads to a situation in which the business model is no longer perceived as a particular (and therefore modifiable) object. Instead, the only traces of the business model that remain visible are the corresponding values. This has far-reaching consequences. Decision makers will tend to reject innovation proposals that contravene the business model implemented

---

[8]The initial application of a business model that already exists in another market sector qualifies in this connection as a business model innovation.

in the enterprise as not viable. The enterprise is thereby rendered incapable of implementing more than one kind of business model.

▶   Decision makers will tend to reject innovation proposals that contravene the business model implemented in the enterprise as *not viable.* The enterprise is thereby rendered incapable of implementing more than one kind of business model.

Gifford Pinchot and Ron Pellman arrive at a similar result from a completely different perspective. In their book *Intrapreneuring in Action* (Pinchot and Pellman 1999), Pinchot and Pellman find the key to innovation in intrapreneurs, i.e. individuals who play the role of entrepreneurs from within an enterprise. Considering our discussion so far, it now becomes clear why intrapreneurs are important. If we strive for disruptive innovations, i.e. business model innovations, we will need to organize our innovation activities in distinct enterprises or autonomous business units. And we will need individuals who operate in the capacity of entrepreneurs to breathe life into such new entities. Pinchot and Pellmann refer to those who act in this capacity within corporate structures as intrapreneurs.

▶   If we strive for disruptive innovations, i.e. business model innovations, we will need to organize our innovation activities in distinct enterprises or autonomous business units. And we will need individuals who operate in the capacity of entrepreneurs to breathe life into such new entities. Those who act in this capacity within corporate structures are referred to as intrapreneurs.

We conclude: Business model innovations are disruptive innovations. They require the formation of an independent enterprise or an autonomous business unit within the enterprise. These will need to be led respectively by an entrepreneur or an intrapreneur.

If an enterprise acquires another enterprise with a different business model, it too will have to refrain from attempting to integrate the acquired enterprise into the existing organizational structure. This is why acquisitions often fail, even of enterprises that already have successful products on the market. In such cases, there may even be a double incompatibility at work, namely, different business models and different enterprise cultures.

In contrast, sustaining innovations leave existing business models unchanged. The corresponding innovation projects are ideally completed in an existing business unit. Moreover, the role of the intrapreneur is not altogether necessary in such cases. Acquired enterprises with the same business model can be integrated into the acquiring enterprise (see Fig. 8.3).

Sustaining Innovation   Disruptive Innovation

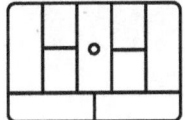   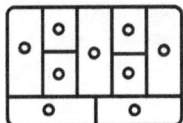

Improved Performance Concept   Often Just Conventional Technology
in an Existing Business Model   in Connection with a New
                                                      Business Model

Fig. 8.3 Sustaining and disruptive innovation in the Osterwalder Business Model Canvas

References

Abell, D. (1993). *Managing with dual strategies: Mastering the present, preempting the future.* New York: Free Press.

Abell, D. (1999). Competing today while preparing for tomorrow. *Sloan Management Review, 40* (Spring), 73–81.

Birkenmeier, B., & Brodbeck, H. (2010). *Wunderwaffe innovation.* Zürich: Orell Füssli.

Chesbrough, H. (2003). *Open innovation.* Boston: Harvard Business School Press.

Christensen, C. M. (1997). *The innovator's dilemma.* New York: Harper Business.

Johnson, M. (2010). *Seizing the white space.* Boston: Harvard Business Press.

Lombriser, R., & Abplanalp, P. A. (2005). *Strategisches management.* Zurich: Versus Verlag.

Osterwalder, A., & Pigneur, Y. (2010). *Business model generation.* New York: Wiley.

Pinchot, G., & Pellman, R. (1999). *Intrapreneuring in action.* San Francisco: Berrett-Koehler Publishers Inc.

Zillner, S., & Krusche, B. (2012). *Systemisches Innovationsmanagement.* Stuttgart: Schäffer-Poeschel Verlag.

Conclusions and Rules Drawn from Part II 9

9.1 Rules for the Enterprise

We have learned that it is necessary to maintain two different enterprise cultures within the same enterprise and to separate the exploration phase of the innovation process from the development part along the cultural fault line.

We have also been forced to revise the standard model of the innovation process by replacing the ideation phase with what we refer to as an early warning system and by also replacing the evaluation and selection phase with what we refer to as the exploration phase. In doing so, we have discovered that exploration represents a crucial and highly complex part of the innovation process, a part that requires a high degree of creativity. The aim of the exploration phase is to transform the ideas that result from ideation into real business opportunities. We have presented a highly iterative process to facilitate the exploration task. We have also found it necessary to introduce a new process phase that we refer to as the transfer phase. The purpose of the transfer phase is to bridge the gap between exploration and development.

Furthermore we have concluded that a capacity for innovation has become increasingly important to enterprises seeking to differentiate themselves from others on the market. This realization makes it necessary to introduce the innovation process as a second main process in addition to the value chain in the Porter enterprise model. And this means that it will be necessary adapt the enterprise structure accordingly.

The following three basic rules apply to the organizational structure of an innovative enterprise:

All illustrations are published with the kind permission of © Heiner Kaufmann, Daniel Huber, and Martin Steinmann. All Rights Reserved.

© Springer International Publishing AG 2017
D. Huber et al., *Bridging the Innovation Gap*, Management for Professionals,
DOI 10.1007/978-3-319-55498-3_9

1. The early warning system and exploration unit are to be given a central position.
2. Business divisions are responsible for their own innovation (sustaining innovation).
3. The exploration unit of the business division is to report to the central innovation unit (dotted-line reporting).

We have proposed a model organizational structure with the following main components. A central innovation unit is to be introduced next to the strategy unit at the level of headquarters. This unit is managed by an innovation officer and also encompasses the early warning system. The two new functions IPR management and venturing are also to be introduced. These can be independent of or integrated into either the strategy unit or the central innovation unit. The business divisions are responsible for their own (sustaining) innovation within their business segment and business model. These innovation units are to report to the central innovation unit at the level of the enterprise (dotted-line reporting).

In the case of small enterprises, the organization of innovation activities takes place in the form of time allocation (workshops and projects).

9.2   Rules for the Strategy

We have also suggested that instead of a one-track enterprise strategy, our enterprise will need a two-track or dual enterprise strategy. This dual strategy is to consist of a today-for-today strategy and a today-for-tomorrow strategy as outlined by Derek Abell. The today-for-today strategy corresponds to the existing enterprise strategy. The today-for-tomorrow strategy corresponds to the existing innovation strategy, which is thereby elevated from the status of a functional partial strategy to the second part of the new dual enterprise strategy and is equally important as the previous today-for-today enterprise strategy.

As a result of this reorganization, enterprise strategy and innovation (particularly exploration) are allowed to enter into a new interactive relationship with one another. On the one hand, strategy is used to steer innovation while on the other hand the results of innovation, especially the results of exploration, may be used to influence strategy development.

We have also suggested that different business models cannot simply be integrated into one single organization. This applies equally to company acquisitions and disruptive innovation projects. These types of projects are to be realized by first establishing subsidiaries or autonomous business units. In such cases, we need individuals to act in the capacity of entrepreneurs or intrapreneurs.

Part III

The Implementation

Rules for Systematic Innovation: The Bern Innovation Model

10

10.1 A Guideline for Systematic Innovation

Of course it would be ideal for an enterprise, if we could give it a failsafe recipe for innovation. However this is obviously an illusion. After all, innovations always are unique and new. However, it is indeed possible to take a systematic approach to innovation. Such approaches to innovation need to be represented in the form of models. This is why we speak of an **innovation model** when we refer to the results of our investigations. To make it easier to present our model as a kind of guideline and for purposes of further discussion, we refer to our model, in particular, as the **Bern Innovation Model**. This is because the investigations that provided the basis for the present book were largely conducted in the Swiss city of Bern.

Our experience working in enterprise settings has shown that it can be extraordinarily helpful to be able to refer to the rules and methods of an innovation model when facing the task of preparing an enterprise for innovation or initiating an innovation project. Given that the innovation model conveys a holistic view of the innovation process, those who use the model for purposes of reference will be led to a holistic solution. This enables one to avoid an overly compartmentalized approach, which turns out to be a very common cause of failure. Measures that are in and of themselves suitable may prove ineffective if one fails to take account of interdependencies in the larger environment. Despite proper measures, the conditions for success have simply not been met.

All illustrations are published with the kind permission of © Heiner Kaufmann, Daniel Huber, and Martin Steinmann. All Rights Reserved.

10.2   The Bern Innovation Model: The New Innovation System

As mentioned, the aim of this fourth part of the book is to present a holistic model of innovation. The elements of innovation described in the first parts are now to be assembled into a complete and coherent model of the innovation process, the Bern Innovation Model. The individual elements will now only be mentioned and positioned in relation to one another. The resulting model can serve as a guideline to successful innovation. It warrants pointing out at this juncture that the innovation activities referred to are not a matter of one-time action, but of continuous efforts.

General Principles of Innovation

In keeping with the definition of innovation we introduced in Chap. 2, according to which innovation always entails the successful deployment of something new (e.g. market success in the case of product innovations and the successful use of a process in the case of process innovations), we can assume that the subject of innovation will always concern the entire enterprise. This is not surprising when we consider the fact that what is at stake is nothing less than securing and shaping the future of the enterprise. Innovation is an eminently strategic matter. As such, it is a matter for the executive management, and the CEO in particular. When it comes to innovation, the CEO will have to perform two tasks: chart the long-term course of the enterprise and secure the implementation of the necessary innovation activities (resource allocation and protection of the relevant innovation activities).

In what follows, we would like to describe the innovation process on the basis of an idealized model, the Bern Innovation Model. This will involve placing the individual elements in relation to one another in a manner that is accessible and intuitive for those wishing to use the model as a reference. To begin, we would like to emphasize the following critical rules:

- The enterprise must *want* to innovate. The members of the executive management and the CEO in particular must be aware of the importance of innovation and the enterprise's intention to innovate.
- It is extraordinarily important in this regard that all of the key decision makers operate according to the same notion of innovation. If this is not the case, the enterprise's innovation efforts will sooner or later derail because any number of these key individuals will have failed to understand one another. For this reason, it is absolutely essential to draft and document the definition of innovation that is to be used within the enterprise from the very outset.
- The definition is to specify that an idea or a project will not qualify as a case of innovation until it is confirmed as such by market success (i.e. real process

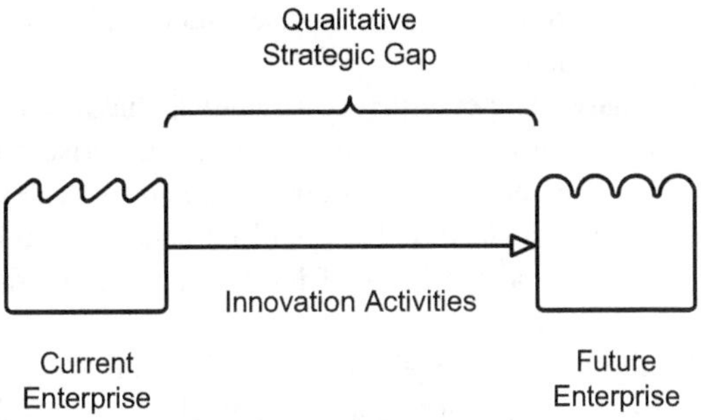

Fig. 10.1 Qualitative
strategic gap

Qualitative
Strategic Gap

Innovation Activities

Current
Enterprise

Future
Enterprise

application in the case of process innovation). If this rule is disregarded, the innovation process is likely to face premature termination (cf. Sect. 2.3).

- In addition to establishing a clear definition of the term innovation, decision makers are to reach a consensus on the significance that innovation is to be accorded within the enterprise. While few are likely to resist its characterization as essential, disputes may arise when it comes to the specific weighting of its significance in the context of resource allocation.

Enterprise Strategy and Innovation Strategy

The second step is to settle all strategic questions:

- First, ascertain what the enterprise is today and what it is to be in the future. This can also be referred to as the **qualitative strategic gap** (see Fig. 10.1). One aim of innovation activities is to close this gap in the form of a today-for-tomorrow strategy (cf. Sect. 8.2).
  (Note: The qualitative strategic gap differs from the **quantitative strategic gap**.[1] It is the task of the actual enterprise strategy (today-for-today strategy) to close this quantitative gap.)
- This will require an inspection of the enterprise strategy. It is important to know what is already contained in the enterprise strategy.
- In addition to examining the documents, attempt to ascertain the views of key players within the enterprise and to determine whether the documents are up to date. This can be handled effectively by conducting interviews with the key players. This especially concerns the CEO, the head of the supervisory board and the head of enterprise strategy.

---

[1]Gabler's dictionary of economic terms defines the strategic gap as the, "... difference between the possible development of an enterprise when presupposing the retention of the current strategic instruments. In contrast to the operational gap, the strategic gap can only be closed with new products, technologies and markets."

http://wirtschaftslexikon.gabler.de/Definition/strategische-luecke.html, as referenced on May 30, 2014.

Depending on the enterprise structure, it may be necessary to confer with other individuals.

- Analyze the enterprise's strategic behavior in the recent past. This will allow you to determine the extent to which enterprise policy and behavior coincide.
- If the enterprise has not documented its strategy or if the documented strategy is no longer up to date, it will be necessary to ascertain and document the basic strategy on the basis of interviews and an analysis of the enterprise's previous behavior.
- Ultimately, it is a matter of being able to answer the two questions mentioned above: "Who are we as an enterprise?" and "Who do want to be as an enterprise?" This qualitative strategic gap is the starting point for every explicit innovative activity and should therefore be set down in writing. It follows that the first document that is to be drafted is the description of the qualitative strategic gap.
- One does not have to work out or revise the entire enterprise strategy to identify the qualitative strategic gap. It will suffice to get in the clear on the vision and mission of the enterprise and the values that are necessary to realize this vision.

We have thereby completed our preliminary assignments relating to the enterprise strategy and can now turn our attention to innovation in particular.

The Innovation Process

To start with, we approach right away the core of the matter: The innovation activities and the innovation process (see Fig. 10.2). Here, our approach resembles that of the previous step involving strategy: We compare our current state to our

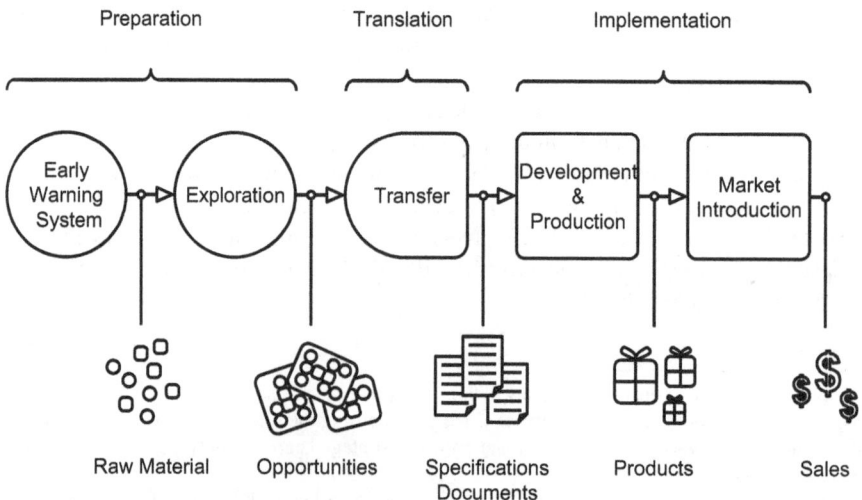

Fig. 10.2 The new innovation process

target state and derive the measures that will be necessary to achieve the target state:

- Locate and examine any existing innovation process documents or preexisting parts thereof.
- Compare the existing process to the model innovation process shown in Sect. 7.3. Position the existing parts of the process in the model process and identify any missing parts.
- Establish the five subprocesses in the model innovation process with complete functionality. This will entail either a complete new setup or supplementing existing parts until they have the complete target functionality.
- It is likely to be the case that the subprocesses exploration and transfer do not exist at all and that the functions of something resembling the early warning system will have to be supplemented to establish a fully functional early warning system.
- The subprocesses of development and market introduction are likely to already exist in fairly well-developed forms. On the other hand, these are likely to be inadequately linked to one another.

Innovation and Enterprise Structure

After we have restructured the components of the innovation process, we will need to make sure that the enterprise's organizational structure also does justice to the new process and its components. It is essentially the case that, as in other contexts, the organizational structure will also have to follow the components of the innovation process, and not the other way around. This is especially the case because the innovation process needs to be fully recognized as a second main enterprise process (see Sect. 5.4).

- As in the case of the processes and the strategy, proceed by taking a close look at the existing organizational structure.
- Then compare the existing organizational structure to the target structure shown in Fig. 10.3.
- As outlined above for processes, start by positioning the existing organizational units in the target structure and supplement these as necessary.
- Make sure to abide by our three organizational rules outlined in Sect. 5.5.
- The elements exploration, early warning system and transfer are likely to be missing from the organizational structure. The functions of IPR management and venturing are probably also missing in most enterprises. Furthermore, just who is responsible for innovation at the level of the executive management will probably also need to be clarified. And finally the responsibilities with regard to innovation will probably need to be clarified between the executive management and the various business units.

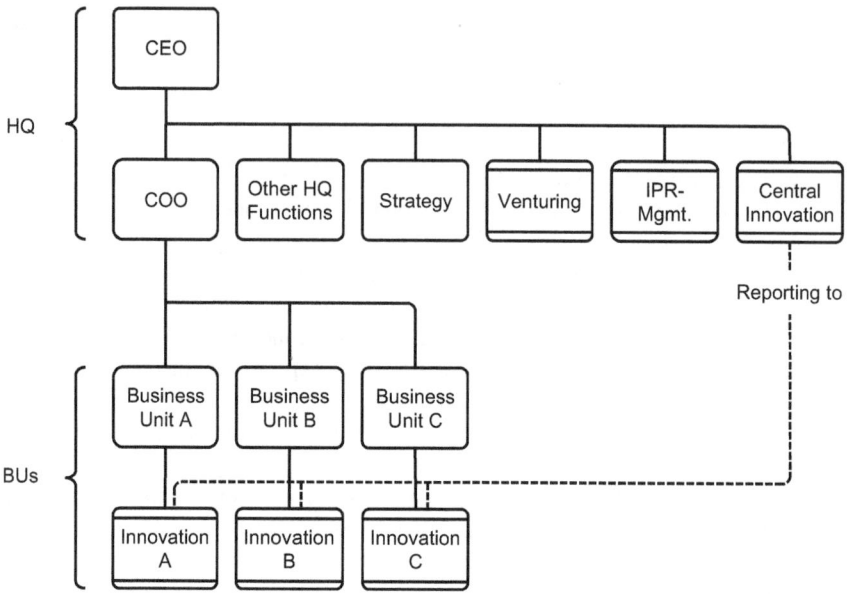

Fig. 10.3 The new organizational structure of the innovative enterprise

- Set up a unit consisting of exploration, early warning system and transfer according to the organizational model outlined in Sect. 6.4 and make sure that this unit outfitted with the robust degree of autonomy that it needs.
- Then determine which of the necessary organizational elements are to take the form of organizational units and which are merely to be rendered in a temporal sense, for instance, as workshops or projects (see Sect. 5.6). Also the size of the organizational units and temporal elements needs to be determined (e.g. number of empolyees, budget, number of required hours or days).

Resources

The necessary financial and human resources are to be provided to ensure a capacity to act. Here it warrants bearing in mind that the desired results can only be achieved if those in charge make clear that they value innovation and are seriously interested in the success of the innovation process.

Financial resources: A single budget item will not be enough. Innovation activities exhibit different characteristics. This is to be taken account of in the context of resource allocation in order to ensure that the innovation efforts both independently of one another and as a whole are allowed to fully develop. The following factors are therefore to be considered in the context of resource allocation:

- The most significant distinction is between one-time and recurring costs. It is important to be aware of items that will involve recurring costs already when first considering the allocation of resources to cover them.
  Projects are often approved for funding in moments of general enthusiasm and in the hopes of scoring quick victories. But one is often not in a position to maintain the funding rhythm over the course of several years. Well-calibrated approaches are important to make sure that the valuable set-up work is not too easily sacrificed in the annual budget allocation discussions.
- Basic services and individual projects: in the context of budget allocation, it is important to distinguish between the cost of making general infrastructure available and the cost of specific projects that will be based on such infrastructure.
- Users and clients: who benefits and can therefore be expected to finance the service? When efforts can be clearly assigned to an existing business segment, then a direct assignment of responsibility is usually clear. If the impact of an activity is less clear or longer term, then it is more likely to be ordered and financed by the overall enterprise. In case of doubt and in order to avoid unwanted managerial influence, it will be better to secure an order from the highest level.

In connection with the individual process steps and roles derived in the Bern Model, this will entail the following in particular:

- Exploration budget: an explicit and appropriate budget must be made available to cover activities relating to exploration and the early warning system. As outlined in Sect. 7.4, this budget should remain stable for the long term. The budget is therefore to be authorized by the executive management.
- Development budget: the budgets for the development and market-introduction phases are to be considered for each individual innovation project. The consideration is to include the costs of the transfer phase. The budget is therefore to be authorized by the division executive management respectively.
- Budgets for the IPR management[2] and venturing functions[3] are to be provided in accordance with their status as central functions. The human resources costs are to be budgeted and funds will also have to be made available for their core activities. The budget is therefore to be authorized by the executive management.

---

[2]The costs of IPR management include the costs of external specialists (e.g. patent lawyers) and fees incurred for IPR protection. Owing to the potentially high costs, the defense of IPRs will usually have to be decided by the executive management on a case-by-case basis.

[3]Funding is to be provided to cover the venturing function. From a certain enterprise size, this is likely to be managed via an internal venturing fund. This will cover both acquisitions and internal startups. The fund will usually only suffice to secure a capacity for action and not for delegating decisions. Decisions concerning individual projects are to continue to be made at the level of the enterprise case by case.

- Project costs associated with enterprise reorganization: as outlined above, improvements in the enterprise's capacity for innovation require a reorganization of the enterprise. These one-time costs are to be estimated and covered. The budget is therefore to be authorized by the executive management.

Many managers equate the allocation of resources for purposes of innovation with the making of investments, which is not quite accurate. While allocating resources to support innovation activities is a matter of investing in the future and while it may indeed take some time before doing so translates into increased sales, such funding should not be regarded as an ad hoc item that can be postponed whenever profits fall. The consequences of doing so may be fatal:

- While enterprises can survive for long stretches on the fruits of their previous efforts, it would be a mistake to assume that no damage has already occurred just because the effects of stagnation won't turn up in the books until later.[4]
- The effects of withdrawing funding will be immediate and not something that will only be felt in the long term! Progress made in recent months or years may essentially be destroyed and current projects will suffer.
- Once you hit the brakes, it will be much more difficult and costly to gather speed again.
- While a complete cessation of funding for innovation activities will be catastrophic, the innovation system will also react allergically to cuts and fluctuation. The lack of steady funding will limit the development of opportunities in the short term and will have disastrous consequences in the long term.

It is therefore important to ensure the availability of appropriate resources for innovation activities. Struggling enterprises, too, will have to retain their capacity for innovation. Even in times of crisis and severely limited resources, there remain smart, innovation-promoting behavioral options.[5]

Human resources: Comprehensive innovation management requires a minimal number of individuals to reach the critical mass necessary for the desired impact. It may also make sense, however, for enterprises to take discreet steps towards this critical mass over a period of time. It often doesn't take a great number of employees to bring early-stage innovation projects to successful completion.

---

[4]CEOs are generally aware of how damaging it can be to merely tread water, to merely process existing orders without new acquisitions while allowing the innovation pipeline to run dry. Nonetheless, the incentives for acting in the long-term interest of the enterprise despite a full-blown crisis are typically weak.

[5]Enterprises that continue to free up resources for innovation in times of austerity send an important signal to their employees. Doing so shows that one believes in the future of the enterprise. The impact of this signal should not be underestimated and can release unsuspected strengths even in individuals under a lot of pressure. That being said, it is less important in such contexts to gain a greater willingness to work for a further bonus than to dispel a depressive and paralyzing work atmosphere. Experienced turnaround managers often emphasize the importance of keeping the prospects for innovation visible in times of crisis so as to keep open a full array of financing options.

The following points are important to bear in mind:

- Not everyone has an aptitude for work as an innovator. While requiring an innovative approach from every employee may send a signal that contributions from all sides are desired, it makes for a rather unrealistic global policy.
- Give careful thought to the skills, experience and character traits that are necessary to reach the established goals. Don't assign those employees who happen to be available at the moment to projects, but those employees who have the right skills.
- Aim for a good mixture. The presence of a number of critical thinkers is likely to be more valuable than complete uniformity. Different people with different backgrounds who work together on innovation projects are likely to reach better outcomes.
- Make sure that employees have the latitude they need to perform. Avoid or mitigate conflicts that derive from hierarchical differences or expected loyalty to original employee departments.
- The recruitment of new employees may be helpful, but is often unnecessary. Dedication is often more important for success than educational background.
- Release the leaders of innovation projects from other obligations. In contrast, it will often suffice to assign other employees to projects on a part-time basis. But here, too, it is important to make sure that they are truly available when they are needed. It is crucial to name names and to honestly appraise the expected impact on standard business operations as early as the rough planning phase. Innovation is not a minor additional assignment that can be lightheartedley added on top of an existing job description. All of the participating employees are to be substantially released from their standard obligations. This includes naming the names of their substitutes.
- The employees who are needed for the innovation process may be valued workhorses in their respective departments. The departments may therefore want to deploy them for their own objectives. Provide the affected departments with appropriate incentives to ensure that these key individuals are truly available and not gradually withdrawn from the innovation project after an initial pledge. This may be the only way to make sure that the department will release such employees for the period in question.
- Conscious make-or-buy decisions will also have to be made in the area of human resources. Instead of promoting existing employees, gaps can also be filled in a targeted manner by recruiting new employees or temporary external consultants. This can effectively lead to the introduction of new ideas and to an increase the professional character of projects, as well as helping one to avoid organizational blindness.

Innovation Management Principles

Finally, we turn to what is the most difficult for enterprises: The separate unit that is responsible for the early warning system, exploration and transfer is to be managed

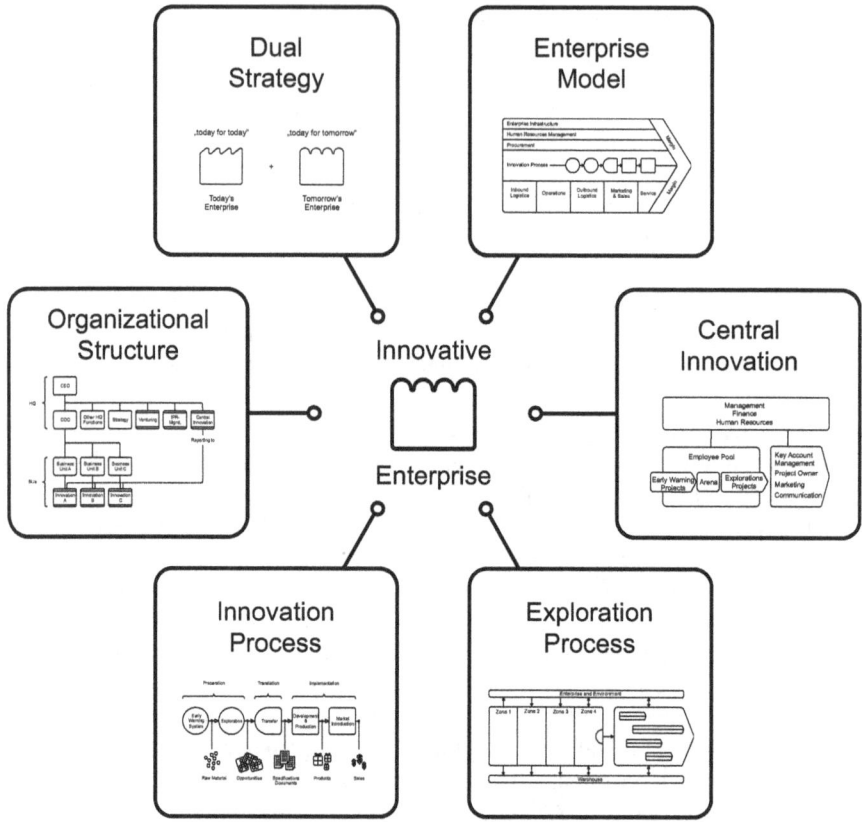

Fig. 10.4 The Innovative Enterprise according to the Bern innovation model

with a set of principles that is strikingly different from that used to manage other enterprise units. Section 5.1 and Chap. 7 offer an extensive account of this different set of management principles.

While this adaptation is bound to be painful for the enterprise, it will determine the extent to which the enterprise is capable of innovation in the future!

Conclusion

The Bern innovation model includes essential new elements at three levels: strategy, process and organization. The model is based on a dual strategy, a comprehensive innovation process and a specific innovation organization. We refer to an enterprise (re)organized in this manner as "Innovative Enterprise". Figure 10.4 offers an overview of the Bern Innovation Model.

Final Remarks

11

11.1 The Innovative Enterprise

We have learned that it is essentially possible to give an enterprise an enhanced capacity for innovation with the Bern Innovation Model. The Bern Innovation Model centers on the management of innovation within enterprises. A capacity for innovation is primarily a matter of the securing the talents of employees with the right personality traits. The crucial factors therefore include employee selection, management and enterprise culture. Of particular relevance in this regard are the aspects of esteem and risk behavior. These requirements also represent an organizational challenge for enterprises because enterprises with an enhanced capacity for innovation are expected to accommodate two essentially opposing management and enterprise cultures at the same time. This unaccustomed and delicate organizational balancing act explains why one rarely meets up with innovative enterprises in the real world. A first step in the right direction will have been taken when both sides of the enterprise, the here-and-now business units and the future-oriented innovation unit, enter into a relationship of mutual recognition and acceptance. When they both regard one another as of equal value for the enterprise, then the enterprise will indeed develop into a sustainably innovative enterprise.

Figure 11.1 offers an illustration of the Innovative Enterprise.

The six steps in the Bern Innovation Model offer a way to an enhanced capacity for innovation. The first step involves an examination of the strategic basis that is necessary for innovation activities. If necessary this basis is to be created anew or an existing basis is to be revised. Building upon this strategic basis, the second step involves the drafting of an innovation strategy. The third and fourth steps involve an adaptation of the enterprise organization, first to accommodate the innovation

---

All illustrations are published with the kind permission of © Heiner Kaufmann, Daniel Huber, and Martin Steinmann. All Rights Reserved.

© Springer International Publishing AG 2017
D. Huber et al., *Bridging the Innovation Gap*, Management for Professionals,
DOI 10.1007/978-3-319-55498-3_11

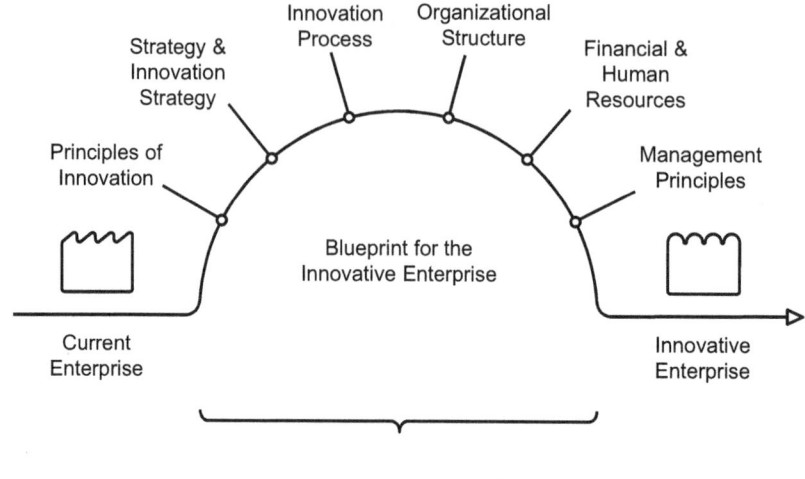

Fig. 11.1   Blueprint of the innovative enterprise

process and second to complete the necessary changes to the organizational structure. The fifth step involves the provision of the necessary financial resources and
human resources. And the sixth step is to define suitable management principles to
govern the two different groups of employees. These are to be implemented in the
form of incentive systems and criteria of promotion.

11.2   A Number of Important Findings

Writing this book has dramatically changed our view of the subject of innovation.
In Table 11.1, we show how our new world view of innovation differs from our
previous view.

Table 11.1 What we thought and what we know now

What we thought	What we know now
Strategy	
The innovation strategy is a substrategy of the enterprise strategy	Innovative enterprises need a dual strategy consisting of a today-for-today strategy and a today-for-tomorrow strategy
Process	
The funnel model of innovation offers an accurate representation of the innovation process	The classic funnel model of innovation is incomplete and doesn't work in the real world. The innovation process involves five distinct phases
First develop the product and then introduce it to the market	Technological exploration and business exploration are to take place simultaneously. They will influence one another in important ways
Functionality and design are crucial for market success	In addition to convincing functionality and design, it is mainly the commercial concept that is crucial to market success
Successful innovation	
High innovation failure rates are normal	Innovation failure tends to occur at specific locations in the process, where it can be dealt with
Ideas	
The value of an innovation idea can be determined in relation to other ideas. The idea-selection process is crucial for successful innovation. Estimating the value of the idea is an essential component in the innovation process	The value of an idea cannot be determined on the basis of the idea alone. The value of an idea depends on the business context into which it is placed. Selecting ideas is not sufficient. The ideas need to be refined in the process of exploration
Tinkering around with ideas is a waste of time. Ideas need to be realized as quickly as possible	In reality, tinkering around with ideas is the key to their refinement and their enhanced commercial value
Management	
Proven methods of effective management also apply to innovation tasks	Standard methods of management are often inappropriate and even counterproductive when it comes to innovation
Careful planning and additional resources will enable one to accelerate the pace of innovation	Breaks and phases of aging are often important for the overall result
Culture	
All employees can be creative and help make the enterprise more innovative	NT personality types are especially suited for achieving success in the creative phases of the innovation process. These individuals are likely to represent a small minority of a given pool of employees. Moreover creativity is only one of many requirements for innovation
An innovation-friendly enterprise culture can be managed from above. Innovation depends solely on the implementation of a culture of innovation	A culture of innovation cannot be directly controlled. It is far more a consequence of creating the right conditions in which it is allowed to grow from the bottom up

11.3   The Answers to Our Research Questions

We can now answer our research questions and their subquestions. The research question we introduced in the box at the end of in Chap. 2 was:

- *What is the best approach to systematic innovation?*

The subquestions were:

- *Why do most innovation projects fail?*
- *What model do enterprises need to follow to achieve successful systematic innovation?*
- *How can this be achieved?*

Let's begin by attempting to provide answers to the subquestions:

- Subquestion 1: Why do most innovation projects fail?

We have shown in the present book that most innovation projects fail for two main reasons:

First, most of the innovation projects that make it to the development phase have not been sufficiently vetted or *explored*. The result is that one doesn't discover exactly what needs to be done until the development phase is underway. This leads to costly and time-consuming iteration loops during the development phase which undermine the cost effectiveness of the desired innovation and can even threaten the existence of the enterprise. We have also discovered why enterprises are so often confronted by this predicament. In short, the commonly used model of the innovation process is incomplete and needs to be expanded through the addition of a new process phase that we refer to as the *exploration phase*.

Second, most innovation projects run into the innovation gap. Innovation projects get terminated between the (new) exploration phase and the development phase on account of cultural differences in the enterprise. While the requirement of accommodating and nurturing two different enterprise cultures within the same enterprise, as well as securing their constructive interaction, is new and demanding, it can be met using the organizational measures, new roles and new principles of management outlined in the present book.

- Subquestion 2: What model do enterprises need to follow to achieve successful systematic innovation?

The Bern Innovation Model provides an answer to this second subquestion. A look at the Bern Innovation Model reveals the strategic and organizational adaptations that are necessary to transform an enterprise into an enterprise with an enhanced capacity for innovation, a so called Innovative Enterprise.

- Subquestion 3: How can this be achieved?

The Bern Innovation Model includes six concrete steps to the creation of the Innovative Enterprise.

By answering the subquestions we have effectively arrived at an answer to our main research question:

- Research question: What is the best approach to systematic innovation?

The best approach to systematic innovation for enterprises is to follow the Bern Innovation Model and the accompanying principles of innovation.

11.4 Conclusion

While innovation can indeed be driven in a systematic manner, doing so involves an adherence to a fundamentally different set of rules than those typically applied in operational enterprise settings. When implementing the generally valid principles of innovation outlined in the present book, it will be important to take account of the relevant unique characteristics of the individual enterprise. Depending on the enterprise or enterprise situation in question, this will then lead to specific implementations that will differ from one another.

Our plan is to continue to work on and refine the Bern Innovation Model and its presentation. We therefore encourage you to remain in contact with us at:

www.innovation-gap.com

The Research Method (Annex) 12

12.1 New Research Methods are Required in the Examination of Innovation

The present annex addresses methodological questions. In the context of carrying out the work we have used as a basis for the present book, it turned out to be surprisingly difficult to package our results and reflections in the form that is typical for publications in the area of economics and business administration. Our examination of the difficulties revealed that they were of a far more fundamental nature than we originally suspected. The subject of our investigations apparently defies meaningful analysis via the statistical methods commonly used in the area of business administration. In the present chapter, we take a more careful look at these methodological issues.

▶ The subject of our investigations defies meaningful analysis via the statistical methods commonly used in the area of business administration.

12.2 General Remarks on the Method

The question about a suitable research method is not trivial in the present case. How can we research something when it is safe to assume that we have either no idea what that something is or a rather skewed understanding of what that something is? In any case, the standard statistical approach is hardly likely to promise success. After all, one can expect that the method will primarily show just how low the chances of success are when it comes to innovation projects, i.e. that innovation as an undertaking doesn't work that well.

All illustrations are published with the kind permission of © Heiner Kaufmann, Daniel Huber, and Martin Steinmann. All Rights Reserved.

© Springer International Publishing AG 2017   161
D. Huber et al., *Bridging the Innovation Gap*, Management for Professionals,
DOI 10.1007/978-3-319-55498-3_12

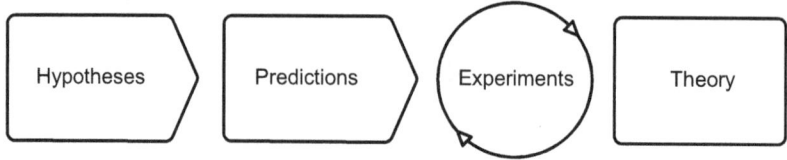

Fig. 12.1 Method of inquiry in the natural sciences

This conclusion, however, doesn't get us anywhere. We wouldn't be able to make any statements about the causes involved or how one might better approach the task of innovation.

On the other hand, this situation, in which the object of research is unknown and is to be researched for precisely that reason, is the classic starting point in the natural sciences. We therefore want to take a closer look in what follows at the typical procedure in research conducted in the natural sciences.

12.3 The Method Applied in the Natural Sciences

If we take a look at how research is conducted in the natural sciences when the phenomena in question are poorly understood, we find the typical approach illustrated in Fig. 12.1.[1]

Observation and reflection are used to develop one or more hypotheses. The process of developing hypotheses is often supported with the use of thought experiments and mathematical methods.

1. Predictions are then made on the basis of a hypothesis. These predictions can be tested using appropriate experiments.
2. The experiments allow one to verify or falsify the predictions.
3. If the predictions are falsified, then the hypothesis or the theory is rejected or modified accordingly.
4. The process is repeated until all of the predictions are verified, at which point the theory qualifies as proven.

Let's now apply this method to our research question and the associated subquestions (see Box at the end of Chap. 2):

- Research question: What is the best approach to systematic innovation?
- Subquestion 1: Why do most innovation projects fail?

---

[1]For instance, see http://de.wikipedia.org/wiki/Theorie, as referenced on May 19, 2014.

Let's first have a look at the subquestion. Here, it is a matter of showing that something doesn't work, which is generally difficult because being able to marshal a proof would presuppose that we know how something works. If we did, we would then be able to demonstrate that certain elements that are a condition for proper functioning are missing and that we can therefore essentially rule out the possibility of proper functioning.[2] Bearing this in mind we are now in a position to answer the research question.

With respect to our research question, we find two possible methodological approaches:

- Research approach 1: In order to be able to show why systematic approaches to innovation fail, we need to find an example of successful systematic innovation. We can examine this functioning innovation system and render it in the form of a model, as a kind of hypothesis. By then comparing the model to other poorly functioning innovation systems, we can identify any elements that are missing from or subverting the failing system. Indeed, the analysis of our functioning system yields a model of a functioning system of systematic innovation, essentially a theory. We therefore can provide a direct answer to our research question.
- Research approach 2: Alternatively, we could also develop hypotheses or theories about how a system of systematic innovation should work on the basis of reflection. By doing so, we remain at the level of a thought experiment and develop our theory on the basis of pure reflection.

Both research approaches suffer from the same weakness because the experiment part of the inquiry remains problematic. The most meaningful prediction for an innovation theory of this sort would be that systematic innovation will now function properly. The question, however, is how such a prediction could be verified. This would require the implementation of the system in a real enterprise and a subsequent measurement of its innovation performance. However, enterprises are highly complex constructs with a vast array of relevant success factors. It is not possible to make precise predictions about their performance. It would therefore be necessary to include many enterprises in the experiment so as to create a reliable statistical basis. Moreover, experience shows that the implementation of an innovation project usually takes several years. Then there is the fact that the activities themselves will be subject to strict confidentiality requirements. Interest on the part of researchers is likely to be met with serious reservations. It therefore seems safe to assume that it would be extremely difficult to demonstrate the effectiveness of our model on the basis of classic experimentation.

However, research approach 1 has a major advantage in that we already have a real-world case of a functioning system of innovation. While this will not suffice for

---

[2]Otherwise, one would have to be able to show that the non-functioning is of a fundamental nature. However, this would make it necessary to prove that something doesn't exist, a condition that simply cannot be met.

an explicit proof, it is a first necessary step in the right direction. It therefore looks like this is the maximum amount of certainty that we can muster short of going to unreasonable lengths. It appears that a genuine statistical proof is out of reach for our research question.

▶   A genuine statistical proof is out of reach for our research question.

12.4   Anecdotal Research

The only viable way of answering our research question appears to involve finding at least one example of the successful implementation of a systematic approach to innovation. A single example will suffice. We don't need a statistical analysis. If we assume that most systematic approaches to innovation fail, an example of a well-functioning system would be an exception and therefore a statistical outlier. We will therefore proceed as follows:

- Look for an example of a well-functioning system—as a unique case and outlier. The aim is therefore to find the very best practice. And best practice is always singular because only one can be the best, in contrast to the statistical average practice.
- Show how the positive (anecdotal) case works by analyzing its properties and using the results of this analysis as a basis for forming our hypothesis.
- Create a model of how an enterprise would have to be structured and managed to be innovative.
- Apply this model in the real world.
- Use the success or failure of the application of our model as evidence of the practical applicability of our model and the soundness of our hypothesis.

We refer to this special research method as **anecdotal research** because it is based on a single, anecdotal case. Anecdotal research is essentially a special case of action research.

▶   Anecdotal research is essentially a special case of action research.

12.5   Our Research Method: Action Research Involving a Real-World Example

This approach will naturally leave us to face other difficulties:

1. How are we to determine whether the innovation in our example has been realized in a systematic manner?
2. And how do we go about finding a case of well-functioning innovation?

We can provide an answer to the first question. The extent to which the innovation efforts of an enterprise have been successful can presumably be best judged by those individuals in the enterprise who are responsible for innovation. This, however, would presuppose that such individuals have a good grasp of our research question and its context. This can usually not be presupposed.

The second question is more difficult and corresponds to the proverbial search for the needle in the haystack. Here, the only thing we can do is hope to find, more or less by chance, a suitable example of a well-functioning systematic approach to innovation. At least we now know what we need to pay attention to while conducting our search. And we also know the method we are to use for our undertaking. This method is action research and involves researching a real, live enterprise.

12.6 Real-World Example

As indicated above, we need a real example of well-functioning, systematic innovation so that we can make progress with our research question using the proposed method. As outlined below, Daniel Huber once had the good fortune of working at a real enterprise as the manager of an innovation unit that offered an example of a long-standing, well-functioning and systematic approach to innovation. This essentially resolves the above-mentioned difficulties. First, we have by chance found a single well-functioning case of innovation. Second, Daniel Huber was a member of the management of the relevant enterprise unit and has at the same time a good understanding of our research question and its context.

In what follows, we now analyze the functioning of the system in question and use our results as a basis for our conclusions.

The case of Swisscom Innovations: As mentioned in our author profiles (see p. xiii), Daniel Huber worked for more than 20 years at Swisscom,[3] where he assumed responsibility for various roles in the enterprise's innovation unit. For several years, he was the deputy head of Swisscom Innovations, the Swisscom Corporation's innovation unit, with around 180 employees. The years in question included a period of a number of years during which the unit's systematic innovation efforts bore fruit.

Our aim in the present chapter is not to offer a systematic description of innovation projects carried out at Swisscom, let alone provide a comprehensive evaluation of these projects. However, it will be necessary to report to a certain extent on this period of innovation so as to demonstrate that it does represent a case of successful, systematic innovation. For reasons of confidentiality, we will naturally refrain from revealing any substantive information on the completed projects. The contents of the projects are not essential for our analysis anyway.

---

[3] Swisscom is the dominant telecommunications provider in Switzerland and was formed out of the former state-owned monopoly PTT. Swisscom's status is therefore comparable to that of Deutsche Telekom in Germany, British Telecom in the UK or AT&T in the US.

In retrospect, various things can be said about this relatively long period of time:

It warrants pointing out that highly professional work was carried out throughout the entire period and this also led to excellent results.

Various phases can be identified when it comes to the relative maturity of the enterprise's capacity for innovation:

Development phase 1 (opportunity-driven): The first phase was characterized by a number of parallel projects that were largely opportunity-driven.

Development phase 2 (formalized 1): The projects were formalized in a second phase and systematically managed after being compiled in a project portfolio. In retrospect, one would have to describe the impact of the completed activities at this point in time on business development for the enterprise as rather isolated. This corresponds to the state of the enterprise's capacity for innovation as one typically encounters in most enterprises today. The corporation itself was more or less on par with the majority of large corporations at the time.

Development phase 3 (formalized 2): In a subsequent phase, the innovation management was improved step by step and generally made more professional. The organizational structure, the processes and the methods were updated in accordance with the latest findings. Systematic searches were conducted for new input and newly published findings were swiftly tested and implemented. As was to be expected, it turned out that not all of the published methods proved to be effective in the context of their practical application. A number of the more well-known theoretical models failed to correspond to reality.[4]

Development phase 4 (innovation laboratory): At the same time, vigorous attempts were made at this point in time to make progress via reflection. In terms of the innovation management, what emerged was a committed trial-and-error culture. In retrospect, one can regard this time as a period of "laboratory experiments" in innovation management. That being said, the "laboratory" was the real world and the subject of examination was a real enterprise. In retrospect, it seems obvious that this is the only possible way of experimenting with innovation management. In the context of this experimental period, we determined that one of the main problems associated with innovation consists of the fact that the innovation process doesn't really work out in reality. This, in turn, was the reason why innovation projects wind up getting terminated prematurely (i.e. before their market readiness). The question naturally arises as to why this is the case. And it is precisely at this point that an example of well-functioning systematic innovation can help us.

And we did get such an example: From one day to the next, the innovation system became consistent and innovation projects began to have a market impact. It was surprising for those directly involved, including Daniel Huber, that this transition to a functioning and complete innovation system took place so

---

[4]For instance, the funnel model of innovation. Please refer to Sect. 7.2 for more.

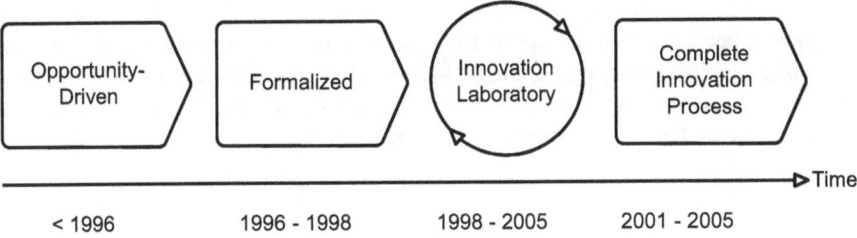

Fig. 12.2 Innovation management at Swisscom innovations

suddenly. Adding to the mystery was the fact that this happened precisely at a time when nothing had been changed on the actual innovation system.

Development phase 5 (established innovation process): A multiple year period in which the Swisscom Corporation consistently demonstrated a capacity for systematic innovation began. Many of Swisscom's more recent products had their origins in this period. All of those directly involved were now very busy and there was a prevailing sense of enthusiasm and hopeful expectation. On the other hand, there was also little opportunity to step back and reflect on what had led to the important changes. Those in the unit were simply happy to see that the innovation machinery worked as they hastened from project to project (see Fig. 12.2).

Development phase 6 (breakup): Just as suddenly as the innovation system began to work, it failed. Once again, everyone was surprised because there were apparently no changes in the innovation system that could have been used to explain the transition to failure. The whole thing remained a mystery for those involved. It wasn't until much later, indeed years after he had left Swisscom that Daniel Huber, aided by the necessary distance to the events themselves, was able to make sense of what had transpired. His assessment is contained in the present book.

12.7 Conclusion

Our initial attempts to answer our research question were met by fundamental difficulties. Given that the usual statistical approach was out of reach, we availed ourselves of the method of anecdotal research, which involves forming a hypothesis on the basis of a single case of effective systematic innovation discovered by chance. We were forced to discover that it wouldn't be possible based on reasonable efforts to present a scientifically sound proof via experimentation. The only thing that now remains is to arrive at a theory that explains the one case of a well-functioning system of innovation.[5] Given our example of a well-functioning system, we should now be able to successfully implement a model of innovation that is

---

[5]In contrast to our previous approaches, this explanation is based on real-world (albeit anecdotal) experience.

based on the theory to other enterprises. The theory developed in the present book has indeed been successfully applied in other enterprises. While we have come up short of scientific proof, this applicability suggests that we are on the right track. Owing to the fundamental difficulties blocking a scientific proof, the only thing that can bolster our theory is its repeated successful application in the real world.

Supporting Organizations

Management Center of the Bern University of Applied Sciences

The Management Center of the Bern University of Applied Sciences in Bern, Switzerland offers a wide range of advanced training programs in management for engineers and other individuals who have had no previous formal instruction in the area of business administration. The Center offers instruction for around 300 students.

As one of seven state universities of applied sciences in Switzerland, Bern University of Applied Sciences encompasses around 6700 students in various fields of study and nearly 3000 employees.

Bern University of Applied Sciences has more than 35 years of experience in continuous education. The Management Center of Bern University of Applied Sciences was the first, and for a long time, the largest, university-level institution to offer advanced training in management. The Center has been a leader in the area of innovation for many years. For instance, it introduced the first full course of studies in the field of innovation more than 15 years ago and also became the first institution of its kind in Switzerland to offer a master's program in innovation.

Otto Wirz Foundation

Headquartered in Bern, Switzerland, the Otto Wirz Foundation pursues the mission of promoting academic and professional training programs as well as students in the field of engineering at the Bern University of Applied Sciences. The Otto Wirz Foundation provided financial support for the present book in the context of one of its promotional projects.

Firmament AG

Founded and based in Bern, Switzerland, Firmament AG offers enterprise consulting services in the area of strategy and innovation. With a special focus on the later, implementation phases of the innovation process, the company helps to develop,

© Springer International Publishing AG 2017
D. Huber et al., *Bridging the Innovation Gap*, Management for Professionals,
DOI 10.1007/978-3-319-55498-3

design and implement user-focused solutions. The company has extensive experience in complex and interdisciplinary business contexts.

In addition to customers in the information and communications technology (ICT) segment, Firmament AG also serves the industrial and administrative sector. Firmament AG is a pioneer of the reverse-acquisition sales method. The company first develops coherent and compelling market opportunities and then approaches suitable market players.

InoBooster

Active and headquartered in Bern, Switzerland since 2012, this network is a specialist for early-phase innovation, especially business exploration. Its customers include enterprises that are interested in becoming more innovative in general or that are in need of an energy boost for specific projects.

InoBooster coined the term business lab and established it as a service for enterprises. The company offers support for transfer activities, including the acquisition and business operationalization of the latest university-based research findings.

InoBooster has successfully implemented many of the approaches outlined in the present book in real-world enterprises.

Glossary

German	English	Explanation (i.e. "used in this Book as" or "newly coined")
Account Manager	Account manager	Person who is responsible for the management of sales and the relationships with particular clients or groups of clients
Action Research	Action research	An approach to research based on participation and action
Anekdotische Forschung	Anecdotal research	A method of action research that is based on a single case or a small number of cases. Anecdotal research offers an alternative means of inquiry when statistical approaches are not feasible
Bausteinkonglomerate	Building block conglomerate	Subsystems comprised of multiple elements or ideas. Such subsystems may themselves be used as components of more complex solutions
Bausteinspeicher	Building block warehouse	Systematic storage of ideas and idea conglomerates for purposes of later use, either alone or in combination with other components
Berner Innovationsmodell	Bern Innovation Model	The new innovation model proposed in the present book
Best Practice	Best practice	Commercial or professional procedures that are accepted as being the best or most effective, and that are therefore used as a benchmark
Business Case	Business case	A systematic account of the reasoning for initiating a new business. In the present book, the term is used to refer to the first rough-and-ready business concept in the exploration phase
Business Exploration	Business exploration	Exploration activities that focus on the business aspects of an innovation proposal, i.e. in contrast to the technical or functional aspects

(continued)

© Springer International Publishing AG 2017
D. Huber et al., *Bridging the Innovation Gap*, Management for Professionals,
DOI 10.1007/978-3-319-55498-3

German	English	Explanation (i.e. "used in this Book as" or "newly coined")
Business Lab	Business lab	A laboratory environment in which various business solutions can be tested in a protected mode without exposure to significant cost and public-image risks. The aim is to acquire an understanding of the behavior of potential customers and competitors, i.e. information that can then be used for the focused, risk-optimized implementation of a business case
Business Plan	Business plan	Formal statement of business goals, reasons they are attainable, and plans for reaching how them
Creaholic SA	Creaholic SA	Innovation enterprise based in Biel, Switzerland (www.creaholic.com)
Dienstleistung	Service	In economics, a service is an immaterial exchange of value. Service is often an economic activity such as labor where the buyer usually does not take exclusive ownership of the product. Using resources, skill, ingenuity, and experience, service providers offer benefits to service consumers. The benefits of a service, if priced, are reflected in the buyer's willingness to pay for it. Service providers participate in an economy without the restrictions of inventory or the need to concern themselves with raw materials (Wikipedia in German, as referenced on May 19, 2014)
Differenziator	Differentiator	That aspect of a product that leads customers to purchase it instead of a competing product
Disruptive Innovation	Disruptive innovation	A disruptive innovation is an innovation that creates a new market and value network and eventually disrupts an existing market and value network, displacing established market leaders and alliances. The term was defined and phenomenon analyzed by Clayton M. Christensen at Harvard University
Entscheidungsvorbereitung	Decision support	Preparation of the information and the documents that are necessary for a decision
Exploration	Exploration	Phase in the innovation process during which initial findings concerning new technical and business solutions are examined in a creative and experimental manner and are combined with or integrated to form entirely new solutions
Explorationsarena	Exploration arena	Figurative arena or playing field for exploration activities. The term alludes to the world of sports to signify that while there are clear rules, our ability to anticipate individual plays is limited

(continued)

German	English	Explanation (i.e. "used in this Book as" or "newly coined")
Firmament AG	Firmament AG	Consulting company based in Bern, Switzerland
Frühwarnsystem	Early warning system	The first phase of the innovation process in which business-relevant new information and developments on markets, in the world of technology and otherwise in the enterprise's environment is identified. The phase also includes ideation
Führungsdreieck	Leadership triangle	An approach to the management of employees that is espoused in the present book. The approach involves combining demands, affirmation and appreciation
Fuzzy Frontend	Fuzzy front end	A term used to refer to the first two phases of the innovation process, i.e. as represented by the traditional funnel model of innovation. The term signifies what many regarded as the necessarily unstructured and uncertain nature of these two phases
Geschäftsmodell	Business model	A business model describes the functioning and logic of a business, specifically the way how it is earning its profits (Wikipedia in German, as referenced on May 19, 2014)
Geschäftsopportunität	Business opportunity	A specific business idea whose implementation has not yet been worked out in detail
Geschäftspotential	Business potential	The potential for business development and growth offered by a set of circumstances
Good Management	Good management	A widely accepted set of rules for ensuring effective management
Hauptdifferenziator	Main differentiator	That property of a product that persuades a majority of customers to purchase it instead of a comparable product
Hauptprozess	Main process	That process that represents the activities of an enterprise that are vital to its success. For instance, Michael Porter refers to the value chain as the main process. Enterprises are advised to align their organizational structures to the main process
Hygienefaktor	Hygiene factor	Product properties whose absence will lead to difficulties, but whose presence will not establish differentiation and increased sales
Innovation	Innovation	A product, method or process that is all of the following: novel, relevant and successful
Innovation Gap	Innovation gap	Gap in the innovation process that appears prior to the development phase and that dooms many innovation proposals
Innovationsmodell	Innovation model	System of elements, including processes and structures, that increases an enterprise's capacity for innovation

(continued)

German	English	Explanation (i.e. "used in this Book as" or "newly coined")
Innovationsstrategie	Innovation strategy	A strategy that describes what an enterprise needs to do to improve its capacity for innovation. Although it is traditionally regarded as a functional substrategy, Abell describes it as an equal part of a dual enterprise strategy
Innovationssystem	Innovation system	The organizational and process structure that enables an enterprise to innovate
Innovatives Unternehmen	Innovative Enterprise	An enterprise with an enhanced capacity for systematic innovation (e.g. according to the Bern Model of Innovation) and that therefore has a sustainable competitive edge
InoBooster AG	InoBooster AG	Innovation consulting company based in Bern, Switzerland. The company specializes in the innovation approaches described in the present book, especially their exploration and business aspects (www.inobooster.com)
Intrapreneur	Intrapreneur	A manager within a company who promotes innovative product development and marketing in an entrepreneurial way
Key Account Manager	Key account manager	A representative in a sales department who is responsible for one or more major clients (key accounts)
Konzeptbaustein	Concept element Concept building block	Element or building block of a solution concept. In contrast to ideas, concept elements are not regarded as complete and finalized elements, but as independent and little attractive parts of a larger whole. While ideas may also be thought of as concept elements, concept elements are usually too inchoate to qualify as ideas
Minimal Viable Product	Minimal viable product	A product that possesses no more than the minimal properties that are necessary for sale (Eric Ries)
Missing Link	Missing link	The missing link in a development chain (originally used to refer to what was regarded as a lack fossil evidence in the theory of evolution)
Mission	Mission	The purpose of an organization. Those activities an enterprise plans to execute to come closer to achieving its vision (cf. mission statement)
Myers–Briggs Type Indicator	Myers–Briggs Type Indicator	Presumably the current most widespread system for categorizing personality types and accompanying test method
Not Invented Here-Effekt	Not-invented-here syndrome	A tendency of social, corporate, or institutional cultures to avoid using or buying already existing ideas, products, research, standards, or knowledge because of their external origins and costs

(continued)

German	English	Explanation (i.e. "used in this Book as" or "newly coined")
NT-Typ	NT personality	Certain type of person who exhibits a tendency to operate on the basis of intuition and rational thought (Myers-Briggs type)
Nutzwertanalyse	Cost-benefit analysis	A systematic approach to estimating the strengths and weaknesses of alternatives. It is a technique that is used to determine options that provide the best approach in terms of benefits in labor, time and cost savings, etc.
Open Innovation	Open innovation	Innovation model that draws elements of the enterprise's environment into the innovation process
Operationelle Exzellenz	Operational excellence	The goal of perfecting the execution of business activities, improving the way business activities are performed
Otto Wirz Stiftung	Otto Wirz Foundation	Foundation dedicated to promoting educational and training programs at Bern University of Applied Sciences
Pflichtenheft	Functional specifications document	Precise description of how the development unit plans to fulfill client (management) requirements
Produkt	Product	Anything that can be offered for sale on a market
Produktinnovation	Product innovation	Innovation via the introduction of a new product, with the processes remaining largely unchanged
Project Factory	Project factory	Organization that is capable of systematically processing a large number of projects, i.e. on an industrial scale, as it were
Prozessinnovation	Process innovation	Innovation via the introduction of new processes, with the products remaining largely unchanged
Qualitativer strategischer Gap	Qualitative strategic gap	Qualitative gap that separates an enterprise from its projected strategic position. The new strategic position is to be reached via innovation
Quantitativer strategischer Gap	Quantitative strategic gap	Earnings gap that arises as a result of the aging process to which existing products are subject. The products are to be renewed or replaced by innovation so as to generate earnings growth
Quick Wins	Quick Wins	Improvements that can be realized directly via the right decisions. No extensive project work or investments are necessary. The term is common in enterprise consulting parlance
SJ-Typ	SJ personality	Certain type of person who exhibits a tendency to operate on the basis of sensing and judging (Myers-Briggs type)

(continued)

German	English	Explanation (i.e. "used in this Book as" or "newly coined")
Start-up Unternehmen	Startup	Newly founded enterprise that is still in search of a business model (Steve Blank)
Sustaining Innovation	Sustaining innovation	Innovation that does not change the rules of the game on the market and whose sale is based on unchanged purchasing criteria (Clayton Christensen)
Swisscom AG	Swisscom AG	The dominant telecommunications provider in Switzerland
Swisscom Innovations	Swisscom Innovations	Innovation unit of Swisscom AG
Technische Exploration	Technical exploration	Procedure for identifying optimal technical solutions for functional requirements
TIME-Industrie	TIME industry	Telecommunications, information, media and entertainment industry
Today for Today Strategie	Today-for-today strategy	Strategy that says what has to be done today to secure today's business (Abell). This strategy corresponds extensively to the classic enterprise strategy
Today for Tomorrow Strategie	Today-for-tomorrow strategy	Strategy that says what has to be done today to secure tomorrow's business (Abell). This strategy corresponds to the innovation strategy
Transfer	Transfer	Important phase in the innovation process marking the transition from exploration activities to implementation activities
Transfermoderator	Transfer moderator	Person responsible for the success of the transition to the innovation implementation phase
Trichtermodell	Innovation funnel Funnel model	Common representation of the innovation process that illustrates the narrowing down of many original ideas and the subsequent development of a few "best" ideas for purposes of innovation
Tütschelen	Tütschelen	Swiss German term roughly equivalent to tinkering and signifying a playful experimentation with and assembly of ideas (objects) with the aim of arriving at a harmonious overall solution. The term derives from the Bernese term "tütschele" which refers to the creative play with building blocks
Unternehmensstrategie	Business strategy	A strategy that reflects the (existing) business of an enterprise and includes proposals for its improvement
Venturing	Venturing	An enterprise unit that is invested with the authority to found new enterprises

(continued)

German	English	Explanation (i.e. "used in this Book as" or "newly coined")
Veredelung	Refinement	A process to which ideas are subjected during the exploration phase of the innovation process. The process leads from initial ideas, which are often revised extensively, to enhanced business opportunities. The business potential may thereby grow by orders of magnitudes
Vision	Vision	An image of a future state that an enterprise can strive to attain. An enterprise's vision can imbue it with a distinct identity
Werte	Values	The values that an enterprise needs to uphold in its internal operation to secure its credibility and enable success

Index

Zeitfracht Medien GmbH
Ferdinand-Jühlke-Straße 7
99095 Erfurt, Deutschland
produktsicherheit@kolibri360.de